I0482166

ACT Math,

Simplified

by Mary Katherine Murphy

www.amazon.com/dp/153366532X/

Printed by CreateSpace, an Amazon.com company.
2020

Why should I buy this book?

ACT Math, Simplified is different from any other math textbook I've ever seen. In it, I have tried to translate complex math problems and complicated vocabulary into simple concepts using language that makes sense to non-mathematicians. This book will teach you how to properly work the math problems (which can help you in your math classes, too), will teach you strategies to figure out multiple choice questions without doing real work, and will also teach you how to work some of the problems on your calculator to save time and brainpower. I want you to understand math, so it isn't scary anymore, and it doesn't seem like a foreign language, either. Come join me. Let's raise your ACT score by simplifying math.

Contents

Chapter 1
Introduction

What is the ACT?

The ACT is a college entrance exam produced by the American College Testing Program. It is designed to measure your knowledge of English, Math, Reading, and Science, as well as your readiness for college-level work. It does NOT measure how smart you are. In general, standardized tests are a good way to measure how well you can take a standardized test.

Why Do I Have to Take It?

Colleges use your ACT scores and your GPA (grade point average) to help determine if you will be accepted to their programs and if you will receive academic scholarships. It can be difficult for colleges to compare GPA's from students at different schools because some schools tend to be easier and to give higher grades for less work while other schools require more work for the same grades. How would you compare a student who took honors classes and received B's to a student who received A's in regular classes? If both students take the ACT, the college has a more fair way to determine which one receives a scholarship.

What Does My Score Mean?

The highest possible score on the ACT is a 36. An average composite score (the final score including all of the sections) is about a 21. In math, average is 19 or 20. You only have to answer about half the questions right to get around 18-21, depending on the particular test. Different Colleges and Universities have different requirements for admission or scholarships. You should talk to your guidance counselor or check the website of your preferred colleges to determine your goal on the ACT.

What's the Format of the ACT?

The ACT is made up of four multiple choice sections and one optional essay. The four sections are always given in the same order: English, Math, Reading, and Science. If you choose to take the Writing section, it will be last.

English	75 questions (5 passages)	45 minutes
Math	60 questions	60 minutes
Reading	40 questions (4 passages)	35 minutes
Science	40 questions (6 or 7 passages)	35 minutes

There is no penalty for wrong answers, so you should never leave a question blank.
The four sections each receive a score between 1 and 36, and then those four scores are averaged to find the composite score.

What's on the Math Section?

Most of the math questions come from Pre-Algebra or Geometry. This chart shows the detailed breakdown of the questions.

Pre-Algebra	14 questions
Elementary Algebra (Algebra I)	10 questions
Intermediate Algebra (Algebra II)	9 questions
Coordinate Geometry	9 questions
Plane Geometry	14 questions
Trigonometry	4 questions

I'm Not Good at Standardized Tests. How Do I Get a Decent Score?

You are not alone. I have met many, many students who are very smart but are not good at standardized tests. I can teach you several strategies to help improve your score and reduce your stress.

-time management

The most important part of any ACT section is the time. You must move quickly through each section. Wear a watch! No, your cell phone doesn't count.

In math, you have 60 minutes to answer 60 questions. That means it's easy to stay on pace. After 10 minutes, you should be at least to number 10. After 20 minutes, be at least to number 20, and so on. In the sections with passages, divide the time by the number of passages. For example, in science, you have about 5 minutes for each passage.

If you don't have time to finish a section, *YOU* choose which questions will be guesses. Skip the hardest reading and science passage(s). Spend your time answering questions that you understand, and just guess on the others. In math, guess on problems that you don't understand or that will take too long. In English, guess on questions that ask about the passage as a whole; they take too long.

-test taking strategies

Answer the easy questions first. In math, the questions generally start easy and get harder. If you're trying to get an average score, spend most of your time on the first half of the test where most of the easy questions are. When you find a hard question that you don't know how to work, guess instead of wasting your time. If you're trying to get a very high score, spend most of your time on the second half of the test. You will need to finish the first half in less than 30 minutes in order to have time to work the harder questions on the second half.

In English, the hardest questions are usually at the end of each passage. If you need to save time, guess on those. If you want a very high score, save time for those.

In reading and science, work passages that make sense to you first. In reading, that's usually the first and third passage, but some students like the second or fourth passage better.

In science, the hardest passage is usually the one with Scientist 1 and Scientist 2 debating a topic. The order of the passages changes with every test, so it could be anywhere in the science section. However, there will only be one of those passages. That passage often doesn't have any graphs or charts, and it can waste too much of your time. Only work that passage if you've already finished all of the others, and you still have time. Also, don't read the science passages. Read the questions and find the answers in the graphs and charts. You rarely have to read the paragraphs.

-practice

Once you have reviewed all of the material, and you know what strategies you plan to use, practice, practice, practice! The more you familiarize yourself with the format of the test, the more you'll be able to use your time wisely and avoid careless mistakes. This book includes math practice tests, and the internet has many more.

-stress management

Finally, don't stress over this test. If you panic on test day, your performance will be disappointing. Relax. It's only a piece of paper. Once you've started college, no one will care what your ACT score was anymore. Most adults don't even remember what their ACT score was. Get a good night's sleep the night before and eat a healthy, filling breakfast the morning of the test.

How Do I Register For the ACT?

Go to ACTstudent.org and create an account. You'll be able to register for the test, see your past scores, and send the scores to colleges. The website will have the latest information on upcoming test dates, locations, and fees. If you don't want to register online, talk to your school's guidance councilor to get a registration packet. If you receive extended time or other accommodations in school, don't forget to request the same for the test, either through the website or your guidance councilor.

What Should I Bring on Test Day?

You will need your admission ticket, a photo ID, number 2 pencils, and your calculator(s).

How Do I Use This Book?

First, take the diagnostic test. The results of it will show you which sections you need to study most. You can choose to skip sections that you already understand if you will be taking the ACT soon. If you have plenty of time before you take the test, you may want to cover every section to ensure you are thoroughly prepared.

Each section in each chapter starts with explanation and examples. Read these carefully if you don't already understand the section. After this are the practice problems. They test your knowledge of the math taught in the section. They are straightforward questions not designed to trick you. After the practice problems are the multiple choice ACT problems. Some of these are trick questions. The ACT designs problems to trick you, and I want you to be familiar with their techniques.

If you read the explanation of a section and are sure you already know how to work the math, you can skip the practice problems. Always work the ACT problems. They may teach you how to avoid a new type of trick question.

At the end of every chapter is a review. Always work the reviews to make sure you haven't missed a concept before you move on to harder sections.

When you finish all of the sections, work another full-length practice test in the allowed time. Wear a watch and check that you're staying on pace. Check your answers and review the questions that you miss. Beware of trick questions!

When test day comes, relax. You're prepared. You know the material, and you've practiced time-saving strategies. You can do this.

Diagnostic Test

1. A function is defined by $f(x)=-2x+5$, and its domain is the set of integers from 0 through 20, inclusive. For how many values of x is $f(x)$ negative?
   A. 17
   B. 18
   C. 19
   D. 20
   E. 21

2. The temperature K in Kelvin is related to the temperature F in Fahrenheit by the equation $K = 0.556F + 255.37$. Which of the following temperatures is closest to 300 Kelvin?
   F. 45°F
   G. 80°F
   H. 122°F
   J. 167°F
   K. 422°F

3. A cupcake recipe requires $1\frac{3}{4}$ cups of sugar to make 24 cupcakes. When the ingredients are increased proportionally to make 60 cupcakes, how many cups of sugar will be required?

   A. $\frac{7}{10}$
   B. $2\frac{5}{8}$
   C. $4\frac{3}{8}$
   D. $4\frac{3}{4}$
   E. $5\frac{1}{4}$

4. Four couples go out to dinner. Each couple will pay for their own meals, but they agree to split the cost of the tip equally among the four. If the total cost of the 8 meals is $276 and the friends leave a 15% tip, What will each couple's portion of the tip be?
   F. $ 5.22
   G. $ 6.90
   H. $ 8.28
   J. $10.35
   K. $18.40

5. Which of the following is not in the solution set of $|2x-4|<12$?
   A. 4
   B. 5
   C. 6
   D. 7
   E. 8

6. What is $\frac{1}{4}$ % of $\frac{8}{9}$?
   F. $\frac{2}{9}$
   G. $\frac{9}{32}$
   H. $\frac{2}{900}$
   J. $\frac{22}{360}$
   K. $\frac{99}{3,200}$

7. The heights of the 5 oak trees in front of the State Capital Building are 77 feet, 83 feet, 82 feet, 81 feet, and 77 feet. What is the mean height of the oak trees?
   A. 65 feet
   B. 77 feet
   C. 80 feet
   D. 81 feet
   E. 83 feet

8. In the figure below, $\overleftrightarrow{AB}$ and $\overleftrightarrow{CD}$ are parallel lines cut by transversals $\overleftrightarrow{AC}$ and $\overleftrightarrow{BD}$. $\overleftrightarrow{AC}$ and $\overleftrightarrow{BD}$ intersect at E. $m\angle ACD = 50°$ and $m\angle BDC = 80°$. What is the measure of $\angle BAC$?

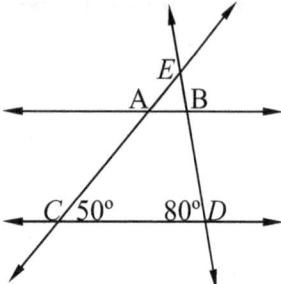

    F. 50°
    G. 80°
    H. 100°
    J. 130°
    K. 180°

9. In Dr. Feldhaus's physics class, Will earned a 74 on one test, an 83 on another test, and 81 on each of the other tests. If Will has an average of exactly 80, and each test is weighted equally, how many tests has Will taken in his physics class?

    A. 3
    B. 4
    C. 5
    D. 6
    E. 7

10. The total cost, c, for the Catahoula Kid's Carnival to rent the fair grounds is determined by $c = 25r + f$, where r is the number of rides the carnival sets up, and f is the fee to rent the fair grounds. Catahoula Kid's Carnival will set up 14 rides this Saturday, and the fee to rent the fairgrounds on a Saturday is $250. They want to be able to cover their costs by selling exactly 200 tickets. How much should Catahoula Kid's Carnival charge for each ticket?

    F. $1.00
    G. $2.20
    H. $3.00
    J. $4.25
    K. $5.00

11. Tabitha rolled a standard six-sided die 30 times. After each roll, she wrote down the digit on the top face. The bar graph below shows her results. What percent of the rolls of the die ended with a 3 on the top of the die?

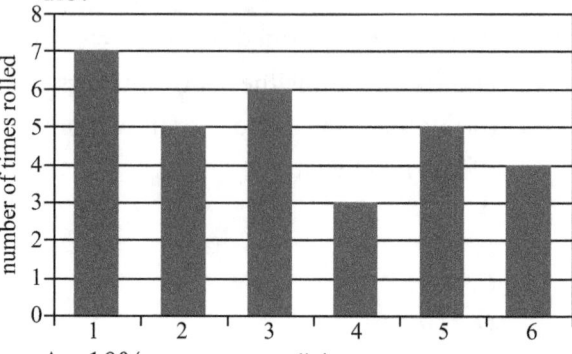

    A. 10%
    B. 13%
    C. 17%
    D. 20%
    E. 23%

12. When $4\frac{6}{8}$ is written as an improper fraction and reduced, the numerator of the fraction is:

    F. 4
    G. 8
    H. 19
    J. 22
    K. 38

13. A bag contains 8 marbles. 4 are red, 3 are green, and 1 is blue. If a marble is drawn at random, returned to the bag, and then a second marble is drawn at random, what is the probability that the first marble is red and the second is green?

    A. $\frac{3}{64}$

    B. $\frac{7}{64}$

    C. $\frac{3}{16}$

    D. $\frac{1}{4}$

    E. $\frac{7}{8}$

Use the following information to answer questions 14 – 16.

Johnny's Pizza is having a special on single topping pizzas with specialty cheese. A customer can order large (14") or medium (12") pizzas with one of 6 different toppings and one of 3 different types of cheese.
The Murphy and LaCroix families each placed orders, as shown in the table below.

Family	Number of pizzas		Cost
	large	medium	
Murphy	6	2	$83.00
LaCroix	4	2	$61.00

14. The cost is the same for any pizza of a given size. What is the cost of a large pizza?
    F. $ 8.50
    G. $ 9.50
    H. $10.00
    J. $10.50
    K. $11.00

15. How many different pizzas can customers order?
    A. 6
    B. 11
    C. 12
    D. 18
    E. 36

16. The Murphy family also purchased 2 orders of breadsticks at $4.00 each and 15 brownies priced at 3 for $2.50. What was the total cost of the Murphy's order without including tax and tip?
    F. $ 89.50
    G. $ 93.50
    H. $ 99.50
    J. $103.50
    K. $128.50

17. The measures of 5 interior angles of a hexagon are 120°, 150°, 110°, 120°, and 145°, respectively. What is the measure of the 6th interior angle?
    A. 75°
    B. 90°
    C. 100°
    D. 120°
    E. 155°

18. At the Riverville High School annual dance-off, the judges will award a cash prize to the winning group. Students may enter in groups of 2, 3, 4, or 5. In order to split the prize money between the dancers in whole-dollar amounts, what is the minimum amount the prize money can be?
    F. $12.00
    G. $15.00
    H. $20.00
    J. $60.00
    K. $80.00

19. If n is a prime number and $x = \sqrt{n}$, which of the following represents a rational number?
    A. $2x$
    B. $\dfrac{x}{2}$
    C. $x+2$
    D. $\sqrt{x}$
    E. x^2

20. A rhombus, RHOM, has vertices that are the midpoints of rectangle ANGL. The lengths of the sides of ANGL are 12 in by 9 in. What is the perimeter of RHOM?
    F. 15 in
    G. 21 in
    H. 30 in
    J. 60 in
    K. 42 in

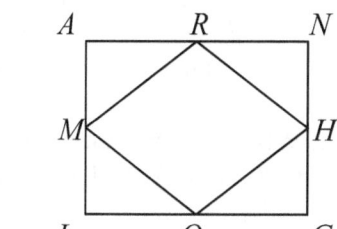

21. In $\triangle ABC$ below, $\sin A = \dfrac{3}{5}$ and the length of $\overline{BC}$ is 7.5 cm. What is the length of $\overline{AC}$?

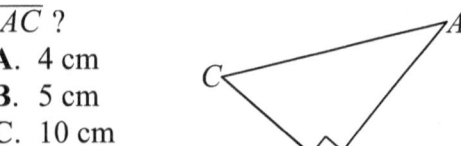

A. 4 cm
B. 5 cm
C. 10 cm
D. 12.5 cm
E. 14.5 cm

22. $3\begin{bmatrix} 1 & 3 \\ 2 & 4 \end{bmatrix} + 2\begin{bmatrix} 2 & -1 \\ 1 & -2 \end{bmatrix} = ?$

F. $\begin{bmatrix} 7 & 7 \\ 8 & 8 \end{bmatrix}$

G. $\begin{bmatrix} 3 & 2 \\ 3 & 2 \end{bmatrix}$

H. $\begin{bmatrix} 3 & 9 \\ 6 & 12 \end{bmatrix}$

J. $\begin{bmatrix} 4 & -2 \\ 2 & -4 \end{bmatrix}$

K. $\begin{bmatrix} 1 & 0 \\ 0 & 1 \end{bmatrix}$

23. If $\sqrt{x} = y$ and $y = 16$, $x = ?$
A. 4
B. 8
C. 32
D. 144
E. 256

24. For what real value of x, if any, is
$\log_{(x+1)}(x^2 + 1) = 2$?
F. -1
G. 0
H. 1
J. 2
K. No such value of x exists.

25. Which of the following shows the number 0.000 000 012 31 written in scientific notation?
A. 1.231×10^{-11}
B. 1.231×10^{-8}
C. 1.231×10^{-7}
D. 1.231×10^{7}
E. 1.231×10^{8}

26. For all positive real numbers x, which of the following expressions is equivalent to
$$\dfrac{\left(\dfrac{x^{16}}{x^5}\right)}{\left(\dfrac{1}{x^4}\right)} ?$$
F. x^4
G. x^7
H. x^{15}
J. x^{17}
K. x^{25}

27. Which of the following is equal to
$\dfrac{5^3}{\sqrt{25}} + \dfrac{\sqrt{72}}{4}$?

A. $5 + \sqrt{3}$
B. $5 + \dfrac{3\sqrt{2}}{2}$
C. 28
D. $25 + \sqrt{3}$
E. $25 + \dfrac{3\sqrt{2}}{2}$

28. What is the length, in coordinate units, of the line segment connecting the points $(2, 2)$ and $(8, -10)$ in the standard (x, y) coordinate plane?
F. $\sqrt{18}$
G. $\sqrt{80}$
H. $\sqrt{108}$
J. $\sqrt{180}$
K. $\sqrt{181}$

29. Triangle $\triangle XYZ$, shown below, is an isosceles triangle. If $\overline{XY} \cong \overline{YZ}$ and $\angle X = 68°$, what is the measure of $\angle Y$?

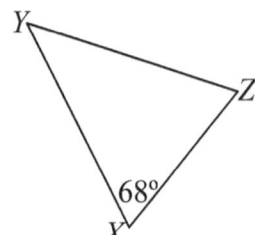

A. 44°
B. 56°
C. 60°
D. 68°
E. 75°

12

30. $\sqrt{98}-\sqrt{50}=?$

   F. $-2\sqrt{2}$
   G. $2\sqrt{2}$
   H. $5\sqrt{2}$
   J. $7\sqrt{2}$
   K. $12\sqrt{2}$

31. $(3x^5+4x^2y^2-6y^2x)-(x^5-3x^2y+5)=?$

   A. $2x^5+4x^2y^2+3x^2y-6y^2x-5$
   B. $2x^5+7x^2y^2-6y^2x-5$
   C. $3x^{10}+12x^4y^2-30y^{2x}$
   D. $4x^5+x^2y^2-6y^{2x}+5$
   E. $4x^5+4x^2y^2-3x^2y-6y^2x+5$

32. Which of the following is a polynomial factor of $2x^2+x-10$?

   F. $(2x-5)$
   G. $(2x-2)$
   H. $(x+5)$
   J. $(2x+2)$
   K. $(2x+5)$

Use the following information to answer questions 33 – 36.

In the standard (x,y) coordinate plane below, X is located at $(0,4)$, Y is located at $(0,1)$, and Z is located at $(4,1)$, forming right triangle $\triangle XYZ$. All lengths are in coordinate units.

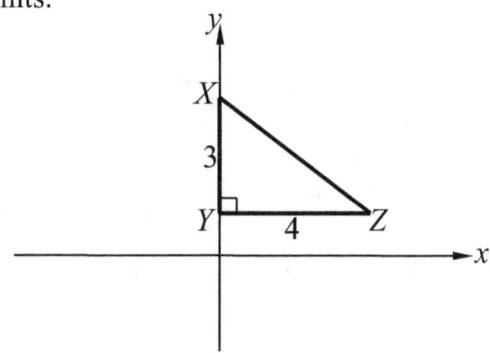

33. What is the midpoint of XZ?

   A. $(2, 2)$
   B. $(0, 2.5)$
   C. $(2, 1)$
   D. $(2, 2.5)$
   E. $(1.5, 2)$

34. What is the slope of XZ ?

   F. $-\dfrac{4}{3}$
   G. $-\dfrac{3}{4}$
   H. -1
   J. $\dfrac{3}{4}$
   K. $\dfrac{4}{3}$

35. Which of the following gives the measure of $\angle Z$?

   A. $\sin^{-1}\left(\dfrac{3}{4}\right)$
   B. $\sin^{-1}\left(\dfrac{4}{3}\right)$
   C. $\cos^{-1}\left(\dfrac{3}{4}\right)$
   D. $\cos^{-1}\left(\dfrac{4}{3}\right)$
   E. $\tan^{-1}\left(\dfrac{3}{4}\right)$

36. If right triangle $\triangle XYZ$ is rotated around the y-axis, a cone will be formed. What will be the length, in coordinate units, of the diameter of the cone?

   F. 3
   G. 4
   H. 5
   J. 6
   K. 8

---

37. The length of a rectangle is 5 inches longer than its width. The area of the rectangle is 24 square inches. Which of the following equations could be solved to find the width of the rectangle?

   A. $w^2=24$
   B. $w(w+5)=24$
   C. $w(w-5)=24$
   D. $2w+2(w+5)=24$
   E. $2w+2(w-5)=24$

38. If k is a constant of proportionality and a, b, c, and d are variables, which of the following equations shows that a varies directly with b and the square root of c and varies inversely with d?

F. $a = \dfrac{kbc}{d}$

G. $a = \dfrac{k b \sqrt{c}}{d}$

H. $a = \dfrac{k b c^2}{d}$

J. $a = \dfrac{k \sqrt{c}}{bd}$

K. $a = k b \sqrt{c} \, d$

39. In the standard (x,y) coordinate plane, which of the following lines is perpendicular to $2x + 3y = 5$?

A. $y = -\dfrac{2}{3} x + 5$

B. $y = -\dfrac{3}{2} x + 5$

C. $y = \dfrac{2}{3} x + 5$

D. $y = \dfrac{3}{2} x + 5$

E. $y = \dfrac{5}{3} x + 2$

40. Given the equation $36x + 12y = 24$, which of the following represents y in terms of x?

F. $-3x + 2$
G. $-3x - 2$
H. $-2x$
J. $3x + 2$
K. $3x - 2$

41. A community garden is a right triangle with the 2 shorter sides having lengths of 9 feet and 12 feet, as shown below. The gardeners will add a decorative fence around all three sides. What length of fence will they need?

A. 21 feet
B. 23 feet
C. 30 feet
D. 36 feet
E. 54 feet

42. The age of Jenny's grandmother exceeds 3 times Jenny's age by 8 years. If j represents Jenny's age, which of the following expressions represents the age of Jenny's grandmother?

F. $j + 8$
G. $j + 11$
H. $j - 8$
J. $3j + 8$
K. $3j - 8$

43. Students studying motion observed a ball rolling at a constant speed along a track. They recorded the distance, d in feet, from the ball to the beginning of the track at one-second intervals starting with $t = 0$. Their results are recorded in the table below.

t	0	1	2	3	4
d	5	8	11	14	17

Which of the following equations correctly represents d in terms of t?

A. $d = t + 5$
B. $d = t + 7$
C. $d = 3t$
D. $d = 3t + 5$
E. $d = 3t + 8$

44. In the triangle below, $\overline{AB} = 3.2$ in and $\overline{BC} = 4.8$ in. $\angle A$ is 110°. Find the measure of $\angle C$.

(Note: For any triangle with vertices A, B, and C opposite sides a, b, and c, respectively, the law of sines states that $\dfrac{\sin A}{a} = \dfrac{\sin B}{b} = \dfrac{\sin C}{c}$ and the law of cosines states that $c^2 = a^2 + b^2 - 2ab\cos C$.)

F. 31.2°
G. 38.8°
H. 42.1°
J. 45.6°
K. 55.0°

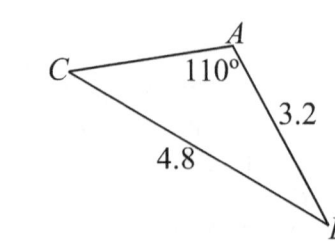

14

45. One of the following inequalities is graphed below in the standard (x,y) coordinate plane. Which one?

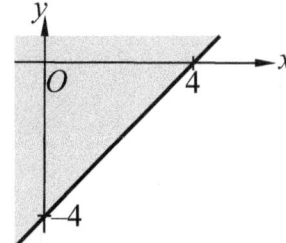

A. $y \geq x+4$
B. $y \geq x-4$
C. $y \leq x-4$
D. $y \geq 4x+4$
E. $y \leq 4x-4$

46. In order to prove that the two triangles shown below are similar by SAS, which of the following pieces of information would you need?

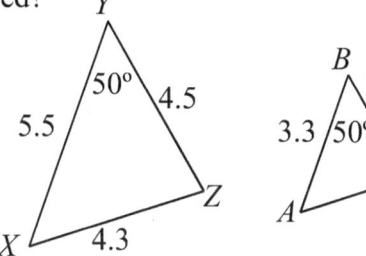

F. $\overline{BC}=2.7$
G. $\overline{BC}=2.3$
H. $\overline{AC}=2.6$
J. $\angle A = 53°$
K. $\angle C = 77°$

47. When graphed in the standard (x,y) coordinate plane, the lines $y=x-5$ and $x=6$ intersect at what point?

A. $(6, -5)$
B. $(6, -1)$
C. $(6, 1)$
D. $(11, -5)$
E. $(11, 6)$

48. What is the solution set to the quadratic equation $x^2-4x+5=0$?

F. $\{1, 5\}$
G. $\{2+i, 2-i\}$
H. $\{-1, -5\}$
J. $\{1+2i, 1-2i\}$
K. $\{-1, 5\}$

49. In a certain board game, the circular spinner is divided into congruent sectors, each with a central angle of 40°. If the diameter of the spinner is 6 inches, what is the arc length of each sector? (Use $\pi = 3.14$)

A. 2.09 in
B. 2.36 in
C. 6.28 in
D. 9.42 in
E. 18.8in

50. In the figure below, all dimensions are in cm and all lines intersect in right angles. What is the area, in cm², of the shaded region?

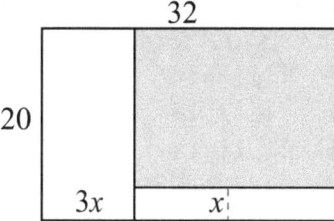

F. $3x^2$
G. $640-3x^2$
H. $640-32x+3x^2$
J. $640-60x+3x^2$
K. $640-92x+3x^2$

51. What is the set of real solutions for $|x|^2-2|x|-3=0$?

A. $\{3\}$
B. $\{-3, 3\}$
C. $\{-1, 3\}$
D. $\{1, 3\}$
E. $\{-3, -1, 1, 3\}$

52. What is the area of the trapezoid below?

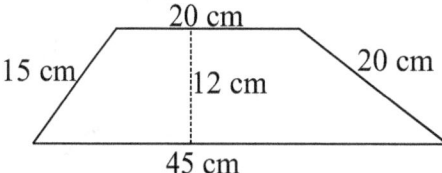

F. 390 cm²
G. 420 cm²
H. 540 cm²
J. 675 cm²
K. 780 cm²

15

53. If $2x^2 - 5x - 12 = 0$, $x = ?$

A. $-\dfrac{3}{2}, -4$

B. $-\dfrac{3}{2}, 4$

C. $-\dfrac{2}{3}, 4$

D. $\dfrac{3}{2}, -4$

E. $\dfrac{3}{2}, 4$

54. The perimeter of a rectangle is 50 meters, and the length of one side is 11 meters. If it can be determined, what are the lengths, in meters, of the other sides?

F. 11, 14, 14
G. 11, 13, 13
H. 11, 11, 14
J. 11, 4.5, 4.5
K. Cannot be determined

55. The graph of $y = \cos x$ is shifted up 3 units, right 0.25π units, and stretched vertically by a factor of 2. Which of the following equations represents this new graph?

A. $y = \cos(2x + 0.25\pi) + 3$
B. $y = \cos(2x - 0.25\pi) + 3$
C. $y = 2\cos(x + 0.25\pi) + 3$
D. $y = 2\cos(x - 0.25\pi) + 3$
E. $y = 2\cos(x - 0.25\pi) - 3$

56. In the standard (x, y) coordinate plane below, a circle has a radius of r coordinate units and passes through the origin, O. The circle's diameter, $\overline{AO}$, lies on the negative x-axis. What are the coordinates of A in terms of r?

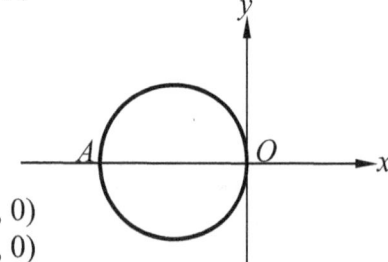

F. $(\ r, 0)$
G. $(-r, 0)$
H. $(\ 2r, 0)$
J. $(-2r, 0)$
K. $(-2\pi r, 0)$

57. The circumference of a circle is 8π units. What is the area, in square units, of the circle?

A. 4π
B. 16π
C. 18π
D. 32π
E. 64π

58. A rectangular swimming pool has dimensions of 15 feet by 25 feet, and a uniform depth of 6 feet. The maximum depth of water that the pool can safely hold is 5 feet. What is the maximum volume of water the pool can hold?

F. 1875 ft^3
G. 2250 ft^3
H. 2400 ft^3
J. 3125 ft^3
K. 11250 ft^3

59. The point $A(5, 3)$ is reflected across the line $x = 8$. What are the coordinates of the point's image, A'?

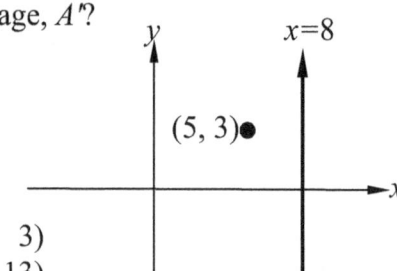

A. $A'(11,\ 3)$
B. $A'(11, 13)$
C. $A'(13, 11)$
D. $A'(15,\ 3)$
E. $A'(15, 13)$

60. A cube has a total surface area of 150 ft². Which of the following expressions gives the area, in square feet, of one face of the cube?

F. $\sqrt[3]{150}$

G. $\sqrt{150}$

H. $\left(\dfrac{150}{6}\right)^2$

J. $\dfrac{150}{6}$

K. $\dfrac{150}{4}$

Answer Key and Action Plan

Question Number	Review Section	Foundations	Pre-Algebra	Elementary Algebra	Intermediate Algebra	Coordinate Geometry	Plane Geometry	Trigonometry
1. B	2.1 Integers 5.1 Solving Inequalities	23			93			
2. G	2.2 Decimals 4.1 Evaluating Expressions and Formulas	26		73				
3. C	2.3 Fractions and Mixed Numbers, 3.1 Ratios, Proportions and Cross-Multiplication	28	49					
4. J	3.2 Percents		51					
5. E	5.2 Absolute Value Equations and Inequalities				95			
6. H	2.3 Fractions and Mixed Numbers, 3.2 Percents	28	51					
7. C	3.3 Mean, Median, and Mode		55					
8. J	7.2 Angles						151	
9. C	3.3 Mean, Median, and Mode		55					
10. H	4.1 Evaluating Expressions and Formulas			73				
11. D	3.2 Percents 3.4 Probability		51 57					
12. H	2.3 Fractions and Mixed Numbers	28						
13. C	3.4 Probability		57					
14. K	5.3 Solving Systems of Equations				96			
15. E	3.5 Organized Counting		59					
16. J	4.1 Evaluating Expressions and Formulas			73				
17. A	7.2 Angles, 7.3 Quadrilaterals and Polygons						151 155	
18. J	2.4 Factors and Prime Numbers	32						
19. E	2.4 Factors and Prime Numbers, 2.5 Square Roots, 4.2 Exponents and Radicals	32 35		75				
20. H	7.3 Quadrilaterals and Polygons						155	
21. D	8.1 Triangle Ratios							189
22. F	5.4 Matrices				98			
23. E	2.5 Square Roots	35						
24. G	5.5 Exponents and Logs				101			
25. B	2.6 Exponents and Scientific Notation	37						
26. H	5.6 Rational and Radical Expressions				104			

17

Question Number	Review Section	Foundations	Pre-Algebra	Elementary Algebra	Intermediate Algebra	Coordinate Geometry	Plane Geometry	Trigonometry
27. E	2.7 Order of Operations, 4.2 Exponents and Radicals	40		75				
28. J	6.2 Distance and Midpoint					123		
29. A	7.4 Triangles						159	
30. G	4.3 Operations with Radicals			78				
31. A	4.4 Polynomials			81				
32. K	4.5 Factoring Polynomials			84				
33. D	6.2 Distance and Midpoint					123		
34. G	6.3 Graphing Linear Equations					126		
35. E	8.1 Triangle Ratios 8.2 Inverse Trigonometric Functions							189 194
36. K	7.5 Right Triangles						162	
37. B	4.6 Solving Quadratic Equations			86				
38. G	5.6 Rational and Radical Expressions				104			
39. D	6.3 Graphing Linear Equations					126		
40. F	3.6 Solving Linear Equations		62					
41. D	7.5 Right Triangles						162	
42. J	3.7 Writing Expressions and Equations		64					
43. D	6.3 Graphing Linear Equations					126		
44. G	8.3 Trigonometric Identities and Laws							196
45. B	6.4 Graphing Systems of Equations and Inequalities					129		
46. F	7.6 Similar Triangles, 7.7 Proof and Logic						165 168	
47. C	6.4 Graphing Systems of Equations and Inequalities					129		
48. G	5.7 Complex Numbers, 5.8 Solving Quadratic Equations and Inequalities				107 109			
49. A	7.8 Circles						170	
50. K	7.9 Area and Perimeter Formulas						174	
51. B	5.2 Absolute Value Equations and Inequalities, 5.8 Solving Quadratic Equations and Inequalities				95 109			
52. F	7.9 Area and Perimeter Formulas						174	
53. B	4.6 Solving Quadratic Equations			86				
54. F	7.9 Area and Perimeter Formulas						174	
55. D	8.4 Trigonometry Graphs							200
56. J	6.5 Graphing Conic Sections					133		
57. B	7.9 Area and Perimeter Formulas						174	
58. F	7.10 3-Dimensional Geometry						177	
59. A	6.6 Transformations					137		
60. J	7.10 3-Dimensional Geometry						177	

Explanations

1. **B**. 18 If the function is negative, it's less than zero. $-2x+5<0$ so $x>2.5$. How many numbers from 0 through 20 are bigger than 2.5? 3 through 20. That's 18 numbers.

2. **G**. 80°F Plug 300 Kelvin in for K. $300=0.556F+255.37$ $F\approx80$

3. **C**. $4\frac{3}{8}$ Use ratios.

$$\frac{1\frac{3}{4}\,\text{sugar}}{24\,\text{cupcakes}}=\frac{x\,\text{sugar}}{60\,\text{cupcakes}}\quad x=4\frac{3}{8}$$

4. **J**. $10.35 The tip will be $.15\times\$276=\$41.40$ Divide that by 4 couples to get $41.40\div4=10.35$.

5. **E**. 8 Split the inequality into $2x-4<12$ and $2x-4>-12$ Solve them to get $-4<x<8$. The only answer choice that is not between –4 and 8 is 8.

6. **H**. $\frac{2}{900}$ $\frac{1}{4}\%=\frac{.25}{100}$, so $\frac{.25}{100}\times\frac{8}{9}=\frac{2}{900}$

7. **C**. 80 feet $\frac{77+83+82+81+77}{5}=80$

8. **J**. 130° Ignore line $\overline{BD}$. $\angle BAC$ and $\angle ACD$ are same-side interior angles, so they are supplementary. $180-50=130$

9. **C**. 5 $\frac{74+83+81x}{2+x}=80$ Simplify and cross multiply. $157+81x=80(2+x)$ $157+81x=160+80x$ $x=3$ so $2+x=5$

10. **H**. $3.00 $c=25r+f$ Plug in $r=14$ and $f=250$. $c=25(14)+250=600$ The total cost is $600. Divided among 200 tickets means $3 per ticket. $600\div200=3$

11. **D**. 20% $\frac{what\ you\ want}{total\ stuff\ you\ can\ get}=\frac{6}{30}$

$\frac{6}{30}=0.2=20\%$

12. **H**. 19 $4\frac{6}{8}=4\frac{3}{4}=\frac{19}{4}$ The numerator is the top number, 19.

13. **C**. $\frac{3}{16}$ $\frac{what\ you\ want}{total\ stuff\ you\ can\ get}$ Since we're drawing a marble twice, we multiply the two probabilities. $\frac{4}{8}\times\frac{3}{8}=\frac{1}{2}\times\frac{3}{8}=\frac{3}{16}$

14. **K**. $11.00 Set up two equations and solve.

$$\begin{array}{l}6l+2m=83\\ -(4l+2m=61)\\ \hline 2l\quad =22\end{array}\text{ So }l=11$$

15. **E**. 18 Size times toppings times cheeses. $2\times6\times3=36$

16. **J**. $103.50 $83+2(4)+\frac{15}{3}(2.5)=103.5$

17. **A**. 75° To find the sum of the interior angles, use the formula $180(n-2)$ with $n=6$ for a hexagon. $180(6-2)=720$. $120+150+110+120+145+x=720$ $x=75$

18. **J**. $60.00 If a number is divisible by 4, it's also divisible by 2. Don't worry about the 2. The smallest number divisible by 3, 4, and 5 is $3\times4\times5=60$

19. **E**. x^2 If n is prime, then $\sqrt{n}$ is irrational. To get a rational number, get rid of the square root. $x^2=(\sqrt{n})^2=n$ If n is prime, it is rational.

20. **H**. 30 in Use one triangle to find the length of one side of the rhombus. Use the Pythagorean theorem. $4.5^2+6^2=x^2$ so $x=7.5$ Multiply by 4 to get the perimeter. $7.5\times4=30$

21. D. 12.5 cm $\sin A = \dfrac{opposite}{adjacent}$

$\sin A = \dfrac{3}{5} = \dfrac{7.5}{AC}$ $AC = 12.5$

22. F. $\begin{bmatrix} 7 & 7 \\ 8 & 8 \end{bmatrix}$ $3\begin{bmatrix} 1 & 3 \\ 2 & 4 \end{bmatrix} + 2\begin{bmatrix} 2 & -1 \\ 1 & -2 \end{bmatrix} = ?$

Distribute the 3 and the 2. $\begin{bmatrix} 3 & 9 \\ 6 & 12 \end{bmatrix} + \begin{bmatrix} 4 & -2 \\ 2 & -4 \end{bmatrix}$

Add. $\begin{bmatrix} 3 & 9 \\ 6 & 12 \end{bmatrix} + \begin{bmatrix} 4 & -2 \\ 2 & -4 \end{bmatrix} = \begin{bmatrix} 7 & 7 \\ 8 & 8 \end{bmatrix}$

23. E. 256   Plug in. $\sqrt{x} = y \rightarrow \sqrt{x} = 16$ Square both sides. $(\sqrt{x})^2 = 16^2 = 256$

24. G. 0 $\log_{(x+1)}(x^2 + 1) = 2$ Rewrite in exponent form. $(x+1)^2 = (x^2 + 1)$ FOIL $x^2 + 2x + 1 = x^2 + 1$ Solve for x. $2x = 0$ so $x = 0$

25. B. 1.231×10^{-8} Move the decimal 8 places to the right. The original number was less than 1, so the exponent is negative.

26. H. x^{15} Reduce the top and the bottom first. $\dfrac{x^{11}}{x^{-4}}$ Now reduce the whole thing, top minus bottom. $11 - (-4) = 15$

27. E. $25 + \dfrac{3\sqrt{2}}{2}$ Simplify the exponents and square roots first. $\dfrac{125}{5} + \dfrac{6\sqrt{2}}{4}$ Now reduce the fractions. $25 + \dfrac{3\sqrt{2}}{2}$

28. J. $\sqrt{180}$ $d = \sqrt{(8-2)^2 + (-10-2)^2}$
$d = \sqrt{(6)^2 + (-12)^2} = \sqrt{36 + 144} = \sqrt{180}$

29. A. 44° $\angle X = \angle Z$ so $68 + 68 + Y = 180$ Solve. $\angle Y = 44°$

30. G. $2\sqrt{2}$ $\sqrt{98} - \sqrt{50} = ?$ Factor tree each square root. $7\sqrt{2} - 5\sqrt{2} = 2\sqrt{2}$

31. A. $2x^5 + 4x^2 y^2 + 3x^2 y - 6y^2 x - 5$ Combine like terms and *only* like terms. The x^5 terms are the only like terms. Don't forget to distribute the negative.

32. K. $(2x+5)$ $2x^2 + x - 10$ factors into $(2x+5)(x-2)$

33. D. (2, 2.5)   Average the x's and average the y's. $\dfrac{0+4}{2} = 2$ $\dfrac{4+1}{2} = 2.5$

34. G. $-\dfrac{3}{4}$ slope $= \dfrac{y_2 - y_1}{x_2 - x_1} = \dfrac{1-4}{4-0} = -\dfrac{3}{4}$

35. E. $\tan^{-1}\left(\dfrac{3}{4}\right)$ $\tan Z = \dfrac{opposite}{adjacent} = \dfrac{3}{4}$

36. K. 8   When the triangle is rotated around the y-axis, it forms a cone with a height of 3 and the base extends 4 on either side of the y-axis. Therefore, the diameter is 8.

37. B. $w(w+5) = 24$ The length is 5 longer than the width, so $l = w+5$. Area is $A = l \cdot w$. Substitute $A = 24$ and $l = w+5$ into the area formula. $24 = (w+5)(w)$

38. G. $a = \dfrac{kb\sqrt{c}}{d}$ The constant always goes on top, along with everything that varies directly, so k, b and the square root of c have to be on top. Anything that varies inversely goes on the bottom, like d.

39. D. $y = \dfrac{3}{2}x + 5$ First rearrange the given line to $y = mx + b$ form. $2x + 3y = 5$ becomes $y = -\dfrac{2}{3}x + \dfrac{5}{3}$. To get a perpendicular line, flip the slope and change the sign. The perpendicular slope is $+\dfrac{3}{2}$. The y-intercept can be anything.

40. F. $-3x + 2$ Rearrange to get y by itself. First move the x term to the other side. $12y = -36x + 24$ Then divide by 12. $y = -3x + 2$

41. D. 36 feet   Use the Pythagorean Theorem to find the missing side. $9^2 + 12^2 = c^2$ so $c = 15$. Add all the sides to find the perimeter. $9 + 12 + 15 = 36$

42. **J.** $3j+8$ Exceeds means it's more than something. Grandmother's age is 3 times Jenny plus 8.

43. **D.** $d=3t+5$ Plug in the first pair of numbers. $t=0$ and $d=5$. **A** and **D** are the only answers that work. Plug in the next pair, $t=1$ and $d=8$. Only answer choice **D** works.

44. **G.** $38.8°$ Plug into the law of sines.

$\dfrac{\sin 110}{4.8}=\dfrac{\sin C}{3.2}$ Cross multiply.

$0.62646=\sin C$ Use inverse sine.

$\sin^{-1}(0.62646)\approx 38.8$

45. **B.** $y\geq x-4$ First find the $y=mx+b$ form of the line. The y-intercept is $b=-4$.

The slope is $m=\dfrac{0-(-4)}{4-0}=1$. Our line is

$y=1x-4$. Plug in a point that is in the shaded region, like (0, 0). The greater than sign makes it true. $0\geq 0-4$

46. **F.** $\overline{BC}=2.7$ The question asks us to use SAS (side-angle-side). We have a pair of sides and an angle, so we need the pair of sides on the other side of the angle. We need $\overline{BC}$ to match up with $\overline{YZ}$.

$\dfrac{5.5}{4.5}=\dfrac{3.3}{\overline{BC}}$ Cross-multiply. $\overline{BC}=2.7$

47. **C.** (6, 1)   The lines are $y=x-5$ and $x=6$. Plug in 6 for x in the other equation. $y=6-5$ so $x=6$ and $y=1$.

48. **G.** $\{2+i, 2-i\}$ Use the quadratic equation.

$x=\dfrac{-b\pm\sqrt{b^2-4ac}}{2a}=\dfrac{4\pm\sqrt{4^2-4(5)}}{2}$

$x=\dfrac{4\pm\sqrt{-4}}{2}=\dfrac{4\pm 2i}{2}=2\pm i$

49. **A.** 2.09 in   Arc length is part of the circumference. $C=\pi d=6\pi$ Use ratios.

$\dfrac{40}{360}=\dfrac{x}{6\pi}$ so $x=\dfrac{2\pi}{3}\approx 2.09$

50. **K.** $640-92x+3x^2$ The length of the shaded rectangle is $(20-x)$ and the width

is $(32-3x)$. Since $A=l\cdot w$, multiply $(20-x)(32-3x)=640-32x-60x+3x^2$. Don't forget to combine like terms.

51. **B.** $\{-3, 3\}$ Factor, using $|x|$ as the variable. $|x|^2-2|x|-3=(|x|-3)(|x|+1)$ so $|x|=3$ or $|x|=-1$. An absolute value can't be negative, so ignore the second solution. If $|x|=3$, then $x=\pm 3$.

52. **F.** 390   For a trapezoid, $A=\dfrac{1}{2}h(b_1+b_2)$.

Plug in. $A=\dfrac{1}{2}(12)(20+45)=390$

53. **B.** $-\dfrac{3}{2}, 4$ Factor $2x^2-5x-12=0$.

$(2x+3)(x-4)=0$ so $x=-\dfrac{3}{2}$ or $x=4$

54. **F.** 11, 14, 14   If the length of one side of a rectangle is 11, then the length of the opposite side is also 11. The remaining 2 sides equal each other. To find perimeter, add the sides. $P=50=11+11+x+x$ $x=14$. The first 11 was given. The other three sides are 11, 14, and 14.

55. **D.** $y=2\cos(x-0.25\pi)+3$ To shift in the x (or y) direction, subtract the shift from the x (or y). Multiply the stretch in front of the function. $(y-3)=2\cdot\cos(x-0.25\pi)$ Move the 3 to the other side.

56. **J.** $(-2r, 0)$ A has an x- coordinate of 2 radii to the left $(-2r)$ and a y-coordinate of 0.

57. **B.** 16π Circumference $= 2\pi r$, so $r = 4$. $A=\pi r^2=\pi\, 4^2=16\pi$

58. **F.** 1875 ft^3 $V=lwh$ Use the height of the water, not the height of the pool. $V=(15)(25)(5)=1875$

59. **A.** $A'(11, 3)$ The point moves 3 units right to reach the line, and then it moves 3 units right again to make the image A'.

60. **J.** $\dfrac{150}{6}$ There are 6 surfaces on a cube.

Divide the surface area by 6.

Chapter 2
Foundations

The skills in this chapter are rarely tested *by themselves* on the ACT. However, nearly every question in the math section uses these principles. The concepts in every other chapter of this book use these building blocks. If you struggle with any of these sections, review them now before you continue. Establishing a strong foundation first will help you understand the more difficult math in later chapters.

Section 2.1
Integers

Integer is just a fancy name for a whole number that's positive, negative, or zero. You need to know the sign rules for adding, subtracting, multiplying, and dividing them even though your calculator can do it for you. The ACT is a timed test, you need to be able to work quickly. Hard problems should be worked on a calculator, but simple addition and multiplication should be worked in your head faster.

Addition and subtraction rules

If the numbers have the same sign, add and keep the sign.	$7 + 5 = 12$ $-7 - 5 = -12$
If the numbers have different signs, subtract and keep the sign of the bigger one.	$7 - 5 = 2$ $-7 + 5 = -2$
"Plus a minus" and "minus a plus" mean the same thing. You can write it either way you prefer.	$7 + (-5) = 2$ $7 - (+5) = 2$
Just like in English, a double negative is a positive.	$7 - (-5) = 7 + (+5) = 12$

If it helps you, you can think of money when you add and subtract integers. Positive numbers mean you have money, and negative numbers mean you owe money.

If you have $7 and you owe your friend $5, what happens? $7 - 5 = 2$
After you pay your friend, you still *have* $2. That's *positive*.

If you have $5 and you owe your friend $7, what happens? $-7 + 5 = -2$
After you pay your friend, you still *owe* $2. That's *negative*.

What if you owe one friend $7 and you owe another friend $5? $-7 - 5 = -12$
All together, you *owe* your friends $12. That's *negative*.

Example 1

Simplify $-5+(6)+(-3)-4-(-1)+8=?$

First, simplify the double negative.
Now, simplify from left to right.

$$-5+(6)+(-3)-4+1+8=?$$
$$-5+6=1$$
$$1+(-3)=-2$$
$$-2-4=-6$$
$$-6+1=-5$$
$$-5+8=3$$

You should be able to quickly work that, one piece at a time, in your head. If you can't, practice, but know that your calculator is there if you need it.

Multiplication and division rules

If the signs are the same, the answer will be positive.
Remember that double negatives make a positive.

$$5 \cdot 2 = 10$$
$$-5(-2) = 10$$
$$10 \div 5 = 2$$
$$-10 \div -5 = 2$$

If the signs are different, the answer will be negative.

$$-5 \cdot 2 = -10$$
$$5(-2) = -10$$
$$10 \div -5 = -2$$
$$-10 \div 5 = -2$$

A positive OR a negative squared is positive.

$$3^2 = (3)(3) = 9$$
$$(-3)^2 = (-3)(-3) = 9$$

Example 2

Simplify $12(-2) \div (-3) \cdot 4 = ?$
Simplify from left to right.

A double negative makes a positive.

$$12(-2) \div (-3) \cdot 4 = ?$$
$$12(-2) = -24$$
$$-24 \div (-3) = 8$$
$$8 \cdot 4 = 64$$

Absolute value

Absolute value means make the number positive. Technically, it is the distance on a number line from that number to zero. Distance is never negative. Think about it; would you ever say that your school was negative five miles from your house? Of course not! It could be five miles east or five miles west, but never negative.

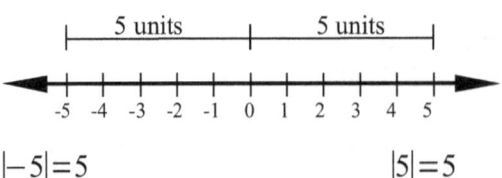

$$|-5|=5 \qquad\qquad |5|=5$$

So, the absolute value of 5 and the absolute value of –5 are both 5.

1. $8-12$

2. $-3+(-9)$

3. $13-(-5)+11+(-1)$

4. $1{,}921+3{,}602$

5. $10{,}621+(-335)$

6. $15 \div |-3| \cdot 5$

7. $8(-5) \cdot (2)(-1) \div 10$

8. $|9||-4||100|$

9. $15 \cdot 12 \div (-10)$

10. $-522 \div 9+(-21)$

11. What is the sum of 11 and –14?

12. Find the product of 15 and –3.

13. What is the difference of –21 and –30?

14. What is the distance on a number line between 82 and –5?

15. Find the quotient of 81 and 3.

16. What is the total of 92, 31, –10, 42, and 3?

1. Mary is 35 years old. Rosa is 12 years younger than Mary. What is the sum of their ages?
   A. 47
   B. 58
   C. 62
   D. 68
   E. 82

2. Sonja needs 1256 yards of yarn to crochet a baby blanket. If each skein of yarn is 314 yards, how many skeins does Sonja need to buy?
   F. 2
   G. 3
   H. 4
   J. 5
   K. 6

3. A roller coaster track starts on the ground and rises 150 feet. It then drops 120 feet around the first curve, rises an additional 75 feet before dropping 80 feet at the second curve. Finally, it rises 33 additional feet before dropping back to the starting location. How high is the track before it returns to the starting location?
   A. 33 feet
   B. 37 feet
   C. 58 feet
   D. 68 feet
   E. 150 feet

4. Mrs. Lang has 27 students in her Kindergarten class. On the first day of school, 18 of them bring a dozen pencils each. If Mrs. Lang collects all of the pencils and divides them evenly among her students, how many pencils will each child receive?
   F. 8
   G. 9
   H. 10
   J. 12
   K. 18

5. Quentin drives on the interstate starting at mile marker 193 and going to mile marker 87. He then drives to mile marker 101 before finally stopping at mile marker 62. How many miles has Quentin driven?
   A. 120
   B. 131
   C. 145
   D. 159
   E. 443

Section 2.2
Decimals

In this book, I assume you will have a calculator when you take the ACT. Therefore, I won't teach you how to multiply 9,675.3128 by 17,305.90405, for example. You should do that problem on a calculator. I want to make sure you remember the things your calculator can't remember for you.

Place value

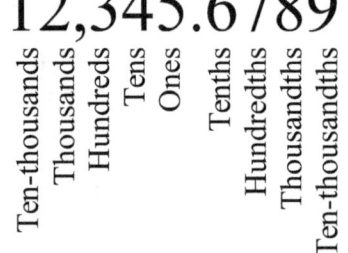

"-ths" means 'after the decimal.'

Place value tells you the value of each digit in a number. For example, in the number below, the value of the *3* is *three hundred*, and the value of the *8* is *eight thousandths*.

Comparing decimals

The symbol > means 'greater than,' and the symbol < means 'less than.' You can remember which is which by thinking about an alligator eating the bigger number.

A line under the greater than or less than symbol means 'or equal to.' The symbol ≥ means 'greater than or equal to,' and the symbol ≤ means 'less than or equal to.'

When comparing multi-digit numbers, use place value starting with the largest values. Compare the largest place value with differing digits.

Example 1
Compare 2.3651 and 2.35821

Line up the decimals to make the place values easier to see.   2.3651
Both numbers have a 2 in the ones place.   2.35821
Both numbers have 3 tenths.
The first has 6 hundredths, but the second has 5 hundredths, so the first is bigger.

$$2.3651 > 2.35821$$

Rounding

When rounding to a certain place value, only the number in that place value has the possibility to change. All of the numbers in front of it will stay the same and all of the numbers behind it will become zero. The number behind the last number decides if the last number will stay the same or go up by one. If the number behind the last one is *5 or greater*, the last number goes up by one. Otherwise, it stays the same.

- 15 rounds up to 20
- 14 rounds down to 10

Example 2

Round 16,444.6988 to the tenths place.

All the numbers in front of tenths stay the same,   16,444.?
and all the numbers after tenths become zero.
The number after tenths (9) decides if the tenths (6) rounds up.
Since $9 \geq 5$, the 6 rounds up to 7.   16,444.7

2.2 Practice Problems: Answers on p. 44

1. What digit is in the hundreds place of 123,456.789?
2. What digit is in the ten-thousandths place of 0.381257?
3. What digit is in the tens place of 159.036? What digit is in the tenths place?
4. Round 57,921.364 to the tens place.
5. Round 361,974 to the nearest thousand.
6. Round 1.268090703 to the nearest whole number.
7. Round 1.499999 to the nearest whole number.
   Compare the numbers. Fill in the blank with >, <, or =.
8. 1.548_____1.562
9. 1,590.617_____1,589.928
10. 6.154_____6.1540

2.2 ACT Problems: Answers on p. 44

1. A new slide for a playground will be 15 feet (180 inches) tall. Each rung of the ladder rises 8.5 inches. Approximately how many rungs will the ladder need?
   A. 15
   B. 18
   C. 21
   D. 24
   E. 27

2. The symbol < makes which of the following choices true?
   F. 6.516_____6.518
   G. 7,320_____7,230
   H. 19.28_____17.28
   J. 12.84_____12.84
   K. 761.4_____758.3

3. Round 834,926.160753 to the nearest ten-thousandth.
   A. 830,000
   B. 834,900
   C. 834,926.0000
   D. 834,926.1607
   E. 834,926.1608

4. Five students competed in the 40-yard dash. Their times are listed below. Who had the fastest time?

Matt	4.76 seconds
Jacob	5.12 seconds
Marcus	4.66 seconds
Drew	4.49 seconds
Aayan	4.39 seconds

   F. Matt
   G. Jacob
   H. Marcus
   J. Drew
   K. Aayan

5. What is the sum of the digits in the tens place and the hundredths place of the number 753.142?
   A. 4
   B. 6
   C. 7
   D. 8
   E. 9

Section 2.3
Fractions and Mixed Numbers

Fractions are another type of problem that should usually be worked on a calculator. Please make sure you know how to add, subtract, multiply, divide, and reduce fractions on *your* calculator, as well as know how to convert between fraction and decimal. However, ACT is very good at making up problems that can't be worked that way. Therefore, you should know how to work these problems on paper just in case you have to.

Reducing
You can reduce a fraction by dividing the **numerator** (top number) and the **denominator** (bottom number) by the same thing. This makes an **equivalent fraction** (equal fraction) to the original. When the ACT asks you to reduce a fraction or to write a fraction in lowest terms, they mean to reduce until you can't make the top and bottom any smaller.

$$\frac{4}{8} \div \frac{4}{4} = \frac{1}{2} \qquad \frac{25}{100} \div \frac{5}{5} = \frac{5}{20} \div \frac{5}{5} = \frac{1}{4} \qquad \frac{14}{21} \div \frac{7}{7} = \frac{2}{3} \qquad \frac{24}{80} \div \frac{4}{4} = \frac{6}{20} \div \frac{2}{2} = \frac{3}{10}$$

Comparing

Comparing fractions with common denominators is easy.

$$\frac{5}{8} > \frac{3}{8}$$

Just compare the numerators.

$$\frac{1}{3} < \frac{2}{3}$$

Another shortcut is comparing fractions with like numerators.

$$\frac{4}{5} > \frac{4}{6}$$

It's the opposite of comparing the denominators.

$$\frac{2}{9} < \frac{2}{3}$$

To compare unlike fractions, we **cross-multiply**.

Multiply the diagonal numbers and write them at the top.

$$14 \quad 15 \qquad 14 < 15$$
$$\frac{2}{5} \bowtie \frac{3}{7} \qquad \frac{2}{5} < \frac{3}{7}$$

$$24 \quad 22 \qquad 24 > 22$$
$$\frac{8}{11} \bowtie \frac{2}{3} \qquad \frac{8}{11} > \frac{2}{3}$$

Example 1
Compare $\dfrac{9}{2} \underline{\quad\quad} \dfrac{11}{3}$

Neither the numerators nor the denominators are the same, so we'll cross multiply.

Multiply the diagonals and write the answers by the tops.

$$27 \quad 22$$
$$\frac{9}{2} \bowtie \frac{11}{3}$$

$27 > 22$

$$\frac{9}{2} > \frac{11}{3}$$

Converting between improper fractions and mixed numbers

An **improper fraction** is a fraction whose numerator is bigger than its denominator. To convert it to a mixed number, divide the top by the bottom. The quotient is your whole number, the remainder is your numerator, and the denominator always stays the same.

Example 2

Convert $\dfrac{31}{4}$ to a mixed number.

4 goes into 31 7 times with 3 left over ($4\times7=28$).

$$\frac{31}{4}=7\frac{3}{4}$$

To convert a mixed number into an improper fraction, multiply the whole number by the denominator and add the numerator. This is your new numerator. The denominator always stays the same.

Example 3

Convert $7\dfrac{3}{4}$ to an improper fraction.

Multiply the whole number by the denominator. $4\times7=28$

Add the numerator. $28+3=31$ $\qquad 7\dfrac{3}{4}=\dfrac{31}{4}$

Adding and subtracting fractions and mixed numbers

When adding or subtracting fractions or mixed numbers, you need to start by finding a common denominator. Remember that you can multiply or divide the top and bottom of a fraction by any number. It's easiest to find the **least common denominator**, the smallest number that both denominators can become.

If you have trouble with mixed numbers, especially with subtraction, you can always use improper fractions instead.

Example 4

$3\dfrac{4}{5}-1\dfrac{1}{2}=?$

First we need a common denominator.

The smallest number that works for 2 and 5 is 10.

Subtract the whole numbers, and subtract the numerators.

$$
\begin{array}{r}
3\dfrac{4}{5}\cdot\dfrac{2}{2}=\;\;3\dfrac{8}{10}\\[2mm]
-\;1\dfrac{1}{2}\cdot\dfrac{5}{5}=-1\dfrac{5}{10}\\[1mm]
\hline
2\dfrac{3}{10}
\end{array}
$$

Multiplying and dividing fractions and mixed numbers

Multiplying fractions is easy. Just multiply the top by the top and the bottom by the bottom. If you have mixed numbers, convert them to improper fractions and multiply. When dividing fractions, flip the second one over and multiply.

Example 5

$$3\frac{4}{5} \div 1\frac{1}{2} = ?$$

First, convert to improper fractions.

$$3\frac{4}{5} = \frac{19}{5}$$
$$1\frac{1}{2} = \frac{3}{2}$$

Now, flip the second one and multiply.

$$\frac{19}{5} \cdot \frac{2}{3} = \frac{19 \cdot 2}{5 \cdot 3} = \frac{38}{15}$$

If you need to, you can convert back to mixed number.

$$\frac{38}{15} = 2\frac{8}{15}$$

Again, I want you to be able to work these problems on paper and on your calculator. If the problem is straightforward, use a calculator if it's faster than you are. However, sometimes ACT comes up with a way to ask their questions so that you can't use a calculator. That's why you need to know how to work these by hand.

2.3 Practice Problems: Answers on p. 44

Compare using >, <, or =

1. $\dfrac{11}{15}$ —— $\dfrac{6}{7}$

2. $2\dfrac{2}{3}$ —— $2\dfrac{7}{10}$

3. $\dfrac{21}{28}$ —— $\dfrac{12}{16}$

4. $\dfrac{8}{9}$ —— $\dfrac{91}{104}$

Simplify. Write your answer in lowest terms.

5. $\dfrac{1}{5} + \dfrac{2}{5} = ?$

6. $\dfrac{1}{2} + \dfrac{2}{3} = ?$

7. $\dfrac{7}{8} - \dfrac{1}{4} = ?$

8. $22\dfrac{5}{6} - 7\dfrac{13}{24} = ?$

9. $1\dfrac{1}{2} - \dfrac{9}{10} = ?$

10. $\dfrac{3}{7} \cdot \dfrac{1}{3} = ?$

11. $2\dfrac{3}{7} \cdot 1\dfrac{1}{6} = ?$

12. $\dfrac{1}{12} \div \dfrac{2}{9} = ?$

13. $5\dfrac{2}{5} \div 3\dfrac{12}{19} = ?$

1. Which of the following correctly lists the fractions $\left\{\dfrac{1}{2}, \dfrac{2}{3}, \dfrac{2}{7}, \dfrac{7}{15}\right\}$ in order from greatest to least?

   A. $\dfrac{2}{3}, \dfrac{1}{2}, \dfrac{7}{15}, \dfrac{2}{7}$

   B. $\dfrac{2}{7}, \dfrac{7}{15}, \dfrac{1}{2}, \dfrac{2}{3}$

   C. $\dfrac{1}{2}, \dfrac{2}{7}, \dfrac{2}{3}, \dfrac{7}{15}$

   D. $\dfrac{1}{2}, \dfrac{2}{3}, \dfrac{2}{7}, \dfrac{7}{15}$

   E. $\dfrac{2}{3}, \dfrac{7}{15}, \dfrac{2}{7}, \dfrac{1}{2}$

2. Maya bought some fruit at the farmers' market. She bought $2\dfrac{2}{5}$ pounds of peaches, $1\dfrac{3}{4}$ pounds of bananas, $3\dfrac{1}{2}$ pounds of apples, and $\dfrac{7}{10}$ pounds of blueberries. How many pounds of fruit did Maya buy?

   F. $6\dfrac{13}{21}$

   G. $7\dfrac{13}{20}$

   H. $7\dfrac{9}{10}$

   J. $8\dfrac{7}{20}$

   K. $8\dfrac{3}{5}$

3. Rebecca's birthday cake is cut into 12 pieces. If Rebecca, Joan, Penny, and Kennedy each have one piece, and John and Layton each have two pieces, what fraction of the cake is left over?

   A. $\dfrac{1}{12}$

   B. $\dfrac{1}{6}$

   C. $\dfrac{1}{4}$

   D. $\dfrac{1}{3}$

   E. $\dfrac{5}{12}$

4. Steve runs $3\dfrac{1}{10}$ miles, but his brother Charlie only runs half as far. How far does Charlie run?

   F. $1\dfrac{1}{2}$

   G. $1\dfrac{1}{3}$

   H. $1\dfrac{11}{20}$

   J. $1\dfrac{1}{20}$

   K. $6\dfrac{1}{5}$

5. $\dfrac{3}{4} \times \dfrac{8}{15} \div \dfrac{11}{3} = ?$

   A. $\dfrac{6}{55}$

   B. $\dfrac{22}{15}$

   C. $\dfrac{14}{30}$

   D. $\dfrac{3}{55}$

   E. $\dfrac{12}{55}$

Section 2.4
Factors and Prime Numbers

Factors are numbers that divide evenly into another number. The factors of 6 are 1, 2, 3, and 6. The factors of 5 are 1 and 5. A **prime number** is a number whose only factors are 1 and itself. 5 is a prime number because its only factors are 1 and 5. A **composite number** has other factors.

Example 1
Find all the factors of 20.

Start with the easiest number: 1. $20 \div 1 = 20$
    1 and 20 are factors.
Try the next number: 2. $20 \div 2 = 10$
    2 and 10 are factors.
Try the next number: 3. $20 \div 3 = 6.6667$
    3 is not a factor.
Try the next number: 4. $20 \div 4 = 5$
    4 and 5 are factors.
We don't have to try 5 because 5 is already a factor.
That's how we know we're finished.

The factors of 20 are 1, 2, 4, 5, 10, and 20.

When you're looking for factors, you don't have to divide by every number in the world. Start with the small numbers and go in order. Once the numbers "wrap around" (the second factor is the same or smaller than the first) you can be sure you found all the factors.

This line "wraps around." →

Factors of 20	
1	20
2	10
4	5
5	4

Divisibility Rules
There are several simple rules that can tell you if a number is a factor without actually dividing to check. These can be very useful with big numbers, especially in school if you don't have a calculator.

Dividing by	Rule
1	Everything can be divided by 1.
2	Even numbers are divisible by 2.
3	Add the digits. If the sum is divisible by 3, the original number is divisible by 3.
4	If the last 2 digits are divisible by 4, the entire number is divisible by 4.
5	If the last digit is 5 or 0, the number is divisible by 5.
6	If the number is divisible by 2 and 3, it's divisible by 6.
7	Use a calculator.
8	If the last 3 digits are divisible by 8, the entire number is divisible by 8.
9	Add the digits. If the sum is 9, the original number is divisible by 9.
10	If the last digit is 0, the number is divisible by 10.

Example 2
Which numbers from 1 through 10 are factors of 123,456,789?

Start at the beginning. 1 is always a factor.
Try 2. 123,456,789 is odd, so 2 is not a factor.
Try 3. Add the digits. $1+2+3+4+5+6+7+8+9=45$ That's still a 2 digit number, so you can add the digits again if you need to. $4+5=9$ 9 is divisible by 3, so 3 is a factor.
Try 4. 89 is not divisible by 4, so 4 is not a factor.
Try 5. Our number ends in 9, so 5 is not a factor.
Try 6. Our number is divisible by 3 but not 2, so 6 is not a factor.
Try 7. Use a calculator. $123,456,789 \div 7 = 17,636,684.\overline{142857}$ 7 is not a factor.
Try 8. 789 is not divisible by 8, so 8 is not a factor. (You may want to use a calculator on that, too.)
Try 9. We already added the digits and got 9, so 9 is a factor.
Try 10. Our number ends in 9, so 10 is not a factor.

The factors of 123,456,789 between 1 and 10 are 1, 3, and 9.

Example 3
Is 53 a prime number or a composite number?

If it's prime, its only factors are 1 and 53. If it has other factors, it's composite. We need to test other factors.
Try 2. 53 is odd, so 2 is not a factor.
Try 3. $5+3=8$ So 3 is not a factor.
We don't need to try 4 or any other even factor because we already know that 53 is odd.
Try 5. 53 doesn't end in 5 or 0, so 5 is not a factor.
Try 7. $53 \div 7 = 7.571$ 7 is not a factor.
We don't have to try 8 or any bigger numbers because 7.571 is less than 8. If there were any bigger factors, we would have already found them.

53 is prime because its only factors are 1 and 53.

Factor trees
If you only need to find the prime factors of a number, you can use a factor tree. Start with any pair of factors of the number, and then break down each of the factors until you are left with only prime numbers.

Example 4
Find the prime factorization of 210.

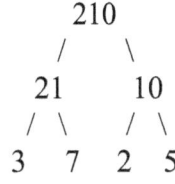

First, find any factors of 210, like 21 and 10.
Next, break 21 and 10 into their factors.
$\qquad 21 = 3 \cdot 7$ and $10 = 2 \cdot 5$
3, 7, 2, and 5 are all prime, so we're finished.

$\qquad 210 = 2 \cdot 3 \cdot 5 \cdot 7$

Greatest Common Factor

The greatest common factor (GCF) of two or more numbers is the biggest number that divides evenly into each of them. The *greatest* common factor will always be *less than* the numbers (or equal to the smallest).

Example 5

What is the greatest common factor of 16 and 24?

Find the factors of the smaller (easier) number.
Starting with the *greatest*, see which are factors
of the other number.

16: 1, 2, 4, 8, 16
$24 \div 16 = 1.5$ not a factor
$24 \div 8 = 3$ 8 is a factor

The GCF of 16 and 24 is 8.

Least Common Multiple

A **multiple** of a number is a number that can be created by multiplying the number by another whole number. The multiples of 3 are 3, 6, 9, 12, 15, etc.

The least common multiple (LCM) of two or more numbers is the smallest number that each of your numbers divides into evenly. The *least* common multiple will always be *greater than* the numbers (or equal to the biggest).

Example 6

What is the least common multiple of 6 and 4?

Find multiples of one of the numbers.
Find multiples of the smaller number until one matches.
12 is in both lists.

6: 6, 12, 18, 24
4: 4, 8, 12

The LCM of 6 and 4 is 12.

2.4 Practice Problems: Answers on p. 45

Find all the factors. Tell if the number
is prime or composite.

1. 13
2. 21
3. 49
4. 55
5. 103
6. 312
7. 555
8. 800
9. 1,000
10. 2,031

Find the prime factorization.

11. 12
12. 30
13. 39
14. 75
15. 90
16. 98
17. 121
18. 800

1. All of the following are prime numbers
   EXCEPT:
   A. 11
   B. 13
   C. 29
   D. 51
   E. 67

2. What is the sum of all the factors of 35?
   F. 12
   G. 13
   H. 47
   J. 48
   K. 56

3. What is the greatest prime factor of 624?
   A. 2
   B. 3
   C. 13
   D. 16
   E. 39

4. What is the smallest nonzero whole number
   that can be divided, separately, by 2, 3, 4, 5,
   and 7 leaving no remainder?
   F. 210
   G. 420
   H. 840
   J. 2070
   K. 5040

5. If a is a positive integer that divides evenly
   into 28 and 42 but does not divide evenly
   into either 12 or 21, what should you get
   when you add the digits in a?
   A. 2
   B. 3
   C. 5
   D. 7
   E. 9

Section 2.5
Square Roots

A square root is the opposite of squared. To find the square root of a number, x, find what number times itself equals x.

1 squared is 1	$1^2=1$	The square root of 1 is 1	$\sqrt{1}=1$
2 squared is 4	$2^2=4$	The square root of 4 is 2	$\sqrt{4}=2$
3 squared is 9	$3^2=9$	The square root of 9 is 3	$\sqrt{9}=3$
4 squared is 16	$4^2=16$	The square root of 16 is 4	$\sqrt{16}=4$

You can even use fractions and decimals.

0.5 squared is 0.25	$0.5^2=0.25$	The square root of 0.25 is 0.5	$\sqrt{0.25}=0.5$
$\frac{1}{2}$ squared is $\frac{1}{4}$	$\left(\frac{1}{2}\right)^2=\frac{1}{4}$	The square root of $\frac{1}{4}$ is $\frac{1}{2}$	$\sqrt{\frac{1}{4}}=\frac{1}{2}$

With a fraction, just take the square root of the top and bottom separately.

$$\sqrt{\frac{4}{9}}=\frac{\sqrt{4}}{\sqrt{9}}=\frac{2}{3}$$

Make sure you know how to find square roots on the calculator that you will use on the ACT.

Example 1

$\sqrt{64} = ?$

If you know your multiplication facts, this is easy. $8 \times 8 = 64$ So $\sqrt{64} = 8$
If you don't remember times tables, use a calculator.

Example 2

If a square has an area of 30 square feet, then the lengths of a side, s, must be between which two consecutive integers?

To find the area of a square using the sides, we square the side length.
To go backwards and find the sides from the area, we find the square root.

We can estimate with whole numbers. $5^2 = 25$ That's too small, but close.
Try something bigger. $6^2 = 36$ That's too big.
We found two consecutive (side-by-side) numbers whose squares are bigger and smaller than 30.

Therefore, $\sqrt{30}$ has to be between 5 and 6. $5 < s < 6$

2.5 Practice Problems: Answers on p. 45

Find the following square roots

1. $\sqrt{25}$

2. $\sqrt{49}$

3. $\sqrt{81}$

4. $\sqrt{144}$

5. $\sqrt{400}$

Estimate the following square roots using consecutive whole numbers.

6. $\sqrt{5}$

7. $\sqrt{10}$

8. $\sqrt{19}$

9. $\sqrt{111}$

10. $\sqrt{125}$

1. The area of a farmer's field is 1.44 square miles. What is the length of each side of the field if the field is in the shape of a square?
   A. 0.22 miles
   B. 0.36 miles
   C. 0.44 miles
   D. 1.20 miles
   E. 1.11 miles

2. $\sqrt{\dfrac{25}{49}} = ?$

   F. $\dfrac{1}{2}$

   G. $\dfrac{1}{4}$

   H. $\dfrac{5}{7}$

   J. $\dfrac{7}{9}$

   K. $\dfrac{5}{49}$

3. Moni is baking cookies for her school's annual bake-off competition. She needs to arrange the cookies on her square dish. She wants an equal number of rows of cookies on each side of her dish for the neatest presentation. If she baked 36 cookies, how many rows of cookies should she arrange on her dish?
   A. 3
   B. 6
   C. 9
   D. 13
   E. 24

4. The area of a square is 52 in^2. The length in inches of each side, x, is between which pair of consecutive integers?
   F. $7 < x < 8$
   G. $8 < x < 9$
   H. $9 < x < 10$
   J. $10 < x < 11$
   K. $11 < x < 12$

Section 2.6
Exponents and Scientific Notation

An exponent means you multiply a number (the base) by itself a certain number of times.

Exponent $\downarrow$
Base $\rightarrow$ $5^3 = 5 \cdot 5 \cdot 5 = 125$ $6^4 = 6 \cdot 6 \cdot 6 \cdot 6 = 1296$ $10^5 = 10 \cdot 10 \cdot 10 \cdot 10 \cdot 10 = 100{,}000$

A fraction can be written as a **negative exponent**.

$$3^{-1} = \frac{1}{3} \qquad\qquad 4^{-2} = \frac{1}{4^2} = \frac{1}{16} \qquad\qquad 10^{-5} = \frac{1}{100{,}000} = 0.00001$$

A square root can also be written as a **fraction exponent**.

$$4^{\frac{1}{2}} = \sqrt[2]{4^1} = 2 \qquad\qquad 9^{\frac{1}{2}} = \sqrt{9} = 3 \qquad\qquad 100^{\frac{1}{2}} = \sqrt{100} = 10$$

Example 1

$$2^4 + 16^{\frac{1}{2}} + 5^{-1} = ?$$

Simplify each term separately.

$$2^4 = 2 \cdot 2 \cdot 2 \cdot 2 = 16$$
$$16^{\frac{1}{2}} = \sqrt{16} = 4$$
$$5^{-1} = \frac{1}{5} = 0.2$$

Now add.

$$16 + 4 + 0.2 = 20.2$$

Scientific notation

To write a number in scientific notation, move the decimal behind the first non-zero digit and multiply by a power of 10. The number of places the decimal moved is your exponent. If your number started out bigger than 1, your exponent will be positive. If your number started out between 0 and 1, your exponent will be negative.

$$2{,}503.1 = 2.503 \times 10^3$$
The decimal moves 3 places, and the number was big, so the exponent is 3.

$$0.0511 = 5.11 \times 10^{-2}$$
The decimal moves 2 places, and the number was smaller than 1, so the exponent is –2.

Example 2

Write 2,501,000 in scientific notation.

First, move the decimal until it is behind the first non-zero digit. We moved the decimal 6 places, and the number was big, so multiply by 10^6.

$$2{,}501{,}000.$$

$$2.501 \times 10^6$$

Remember that you can drop the zeros at the end behind the decimal.

Example 3

Write 4.972×10^{-4} in decimal form.

An exponent of –4 means to make the number smaller, so move the decimal to the left. Fill in the blanks with zeros.

4.972

0.0004972

2.6 Practice Problems: Answers on p. 45-46

Write as a whole number or fraction.

1. 8^3
2. 5^4
3. 10^7
4. 11^{-1}
5. 4^{-2}
6. 10^{-5}
7. $25^{\frac{1}{2}}$
8. $100^{\frac{1}{2}}$
9. $4^{-\frac{1}{2}}$

Write in scientific notation

10. 62.022
11. 32,670
12. 0.000001124

Write in decimal form

13. 6.821×10^2
14. 3.005×10^{-4}
15. 5.599×10^6

2.6 ACT Problems: Answers on p. 46

1. The speed of light is 300,000,000 m/s. Which of the following shows the speed of light written in scientific notation?
   A. 3×10^{-9} m/s
   B. 3×10^{-8} m/s
   C. 3×10^3 m/s
   D. 3×10^8 m/s
   E. 3×10^9 m/s

2. What is the approximate value of $24^{\frac{1}{2}}$?
   F. 5
   G. 6
   H. 8
   J. 12
   K. 18

3. A number multiplied by itself 5 times is 1024. What is the value of the number?
   A. 3
   B. 4
   C. 5
   D. 6
   E. 7

4. The OSHA peak permissible limit of hydrogen sulfide in air is 50 parts per million. Express this concentration in scientific notation.
   F. 5×10^1
   G. 50×10^{-4}
   H. 50×10^{-5}
   J. 5×10^{-4}
   K. 5×10^{-5}

5. $(2^3)(3^{-2}) = ?$
   A. $\dfrac{8}{9}$
   B. 1
   C. 5
   D. 6
   E. 72

Section 2.7
Order of Operations

When working a multi-step problem, you must work the steps in a particular order.

$$\underset{\substack{\text{Parentheses}}}{P}\ \underset{\substack{\text{Exponents}}}{E}\ \underset{\substack{\text{Multiplication} \\ \text{and} \\ \text{Division}}}{M}\ \underset{}{D}\ \underset{\substack{\text{Addition} \\ \text{and} \\ \text{Subtraction}}}{A}\ \underset{}{S}$$

You can also remember this order with **P**lease **E**xcuse **M**y **D**ear **A**unt **S**ally.

P – The first thing you have to do is work anything that is inside parentheses, brackets, absolute value, or other grouping symbols.

E – Next, calculate any numbers with exponents.

M&D – Then work multiplication and division from left to right.

A&S – Finally, work addition and subtraction from left to right.

Example 1
$1+5\times3-4\div2=?$

We have no parentheses and no exponents,	$1+5\times3-4\div2$
so next comes multiplication and division.	$\diagdown\diagup\ \ \diagdown\diagup$
Work from left to right.	$1+\ \ 15\ \ -\ \ 2$
Finally comes addition and subtraction	$\diagdown\ \diagup$
from left to right.	$16\ \ -\ \ 2$
	14

Example 2
$2(5-1)+(3\times4)\div6=?$

Even though subtraction is usually one of the last steps,	$2(5-1)+(3\times4)\div6$
we have to work it first because it's in parentheses.	$2\ (\ 4\ )\ +\ (\ 12\ )\div6$
Multiply and divide.	$8\ +\ 2$
Add.	10

Example 3
$3^2\times(2+3)^2+(9-5)^2\div8-1=?$

Start with parentheses.	$3^2\times(2+3)^2+(9-5)^2\div8$
	$3^2\times(5)^2+(4)^2\div8-1$
Exponents	$9\times25+16\div8-1$
Multiply and divide	$225+2-1$
Add and subtract	226

Calculate each result.

1. $10 \times 8 - 7$

2. $6 + 4 \div 2$

3. $3(13 - 7) \div (3 - 1)$

4. $5 + (11 \times 4) - (3 + 2)$

5. $(15 - 9)^2 + 5(3) - 14$

6. $(1 + 3 \times 4 \div 6)^3$

7. $2^3 + 2^2 + 2$

8. $16 \div (9 \times 1 - 7)^2 + 3(6 \div 2)$

9. $(8 - 7 \times 3) - 2(-16 \div 2)$

10. $(3 + 2 \times 4^2) \div (11 - 18) + (3^3 - 2 \times 11)$

2.7 ACT Problems: Answers on p. 46

1. Which of the following expressions has the largest result?
   A. $3^2 \times 4 - 1$
   B. $3^2 \times (4 - 1)$
   C. $(3 \times 4)^2 - 1$
   D. $(3 \times 4 - 1)^2$
   E. All of the expressions are equal.

2. $|3(-5) + 4(6)| = ?$
   F. -9
   G. 9
   H. 18
   J. 39
   K. 114

3. $|8 - 5| - |3 - 10| = ?$
   A. -4
   B. 0
   C. 4
   D. 10
   E. 26

4. $(3 \times 2^2)^2 + (6 - 10) \div 2 = ?$
   F. 10
   G. 70
   H. 142
   J. 646
   K. 1294

5. $\dfrac{(-2 + 7 \times 5)}{(5 + 2 \times 3)} = ?$
   A. $\dfrac{11}{7}$
   B. $\dfrac{40}{21}$
   C. $\dfrac{25}{11}$
   D. 3
   E. 11

Foundations Review

1. A fence encloses a square area of 100 square yards. What is the total length of fence that surrounds the area?
   A. 10 yards
   B. 20 yards
   C. 25 yards
   D. 30 yards
   E. 40 yards

2. The symbol > makes which of the following choices true?
   F. 0.0024_____0.0240
   G. 12.911_____12.899
   H. 6.1015_____6.1105
   J. 9.8108_____9.9019
   K. 102.02_____1020.1

3. The distance from Natchez to Natchitoches is 132 miles. The distance from Natchitoches to Nacogdoches is 111 miles. The distance from Nacogdoches to Natchez is 233 miles. Jeffrey drives from Nacogdoches to Natchitoches and then to Nachez. How far does he drive?
   A. 233 miles
   B. 243 miles
   C. 344 miles
   D. 365 miles
   E. 476 miles

4. A chemistry experiment requires 6.042×10^{-5} L of a hydrochloric acid solution. What is this amount expressed as a decimal?
   F.   0.000006042L
   G.   0.00006042L
   H.   0.6042L
   J.   6,042L
   K. 60,420L

5. Blair, Jonathan, and Lucy went Trick-or Treating for Halloween. Blair scored $2\frac{3}{8}$ pounds of candy, Jonathan scored $3\frac{1}{4}$ pounds of candy, and Lucy scored $1\frac{15}{16}$ pounds of candy. How many total pounds did the three children collect?

   A. $6\frac{5}{16}$

   B. $6\frac{19}{28}$

   C. $7\frac{3}{16}$

   D. $7\frac{3}{8}$

   E. $7\frac{9}{16}$

6. Find the sum of the complete list of factors of 63.
   F. 40
   G. 80
   H. 88
   J. 103
   K. 104

7. What is the product of 12, −3, and −5?
   A.   180
   B.   27
   C.   4
   D.   −45
   E. −180

42

8. $\left|5(-5)+3(8+1)\right|=?$

    F. −52

    G. −2

    H. 2

    J. 52

    K. 175

9. What is the greatest common factor of 48 and 36?

    A. 2

    B. 6

    C. 12

    D. 18

    E. 36

Answers on p. 46-47

10. $2(9-3)^2+\dfrac{11-(-7)}{3}=?$

    F. 6

    G. $73\dfrac{1}{3}$

    H. 78

    J. 81

    K. 150

Answers and Explanations

Section 2.1 Integers
Practice Problems:
1. −4
2. −12
3. 28
4. 5,523
5. 10,286
6. 25
7. 8
8. 3,600
9. −18
10. −79
11. −3
12. −45
13. 9
14. 87
15. 27
16. 158

2.1 ACT Problems:
1. **B**. 58 First, how old is Rosa? $35-12=23$ Now, find the sum. $35+23=58$
2. **H**. 4 We need to divide. $1256\div314=4$
3. **C**. 58 Remember that up is positive and down is negative. $150-120+75-80+33=58$
4. **F**. 8 Only 18 students bring pencils, so multiply 18 and 12 first. Then divide by 27. $18\cdot12\div27=8$
5. **D**. 159 Find the distance for each leg of the trip separately. Remember that distance is positive. $|87-193|=106$ $101-87=14$ $|62-101|=39$ $106+14+39=159$

Section 2.2 Decimals
Practice Problems:
1. 4
2. 2
3. 5, 0
4. 57,920
5. 362,000
6. 1
7. 1
8. <
9. >
10. =

2.2 ACT Problems:
1. **C**. 21 Use inches divided by inches, not feet divided by inches. $180\div8.5=21.176$ You can't have part of a ladder rung, so round to the nearest whole number, 21.
2. **F**. 6.516____6.518 Use the only digit that's different. $6<8$
3. **E**. 834,926.1608 In 834,926.160753, the 7 is in the ten-thousandths place. The 5 behind it makes it round up to 8.
4. **K**. Aayan Aayan's time of 4.39 seconds is less than all of the other times.
5. **E**. 9 The 5 is in the tens place, and the 4 is in the hundredths place. $5+4=9$

Section 2.3 Fractions and Mixed Numbers
Practice Problems:
1. <
2. <
3. =
4. >
5. $\dfrac{3}{5}$
6. $1\dfrac{1}{6}$
7. $\dfrac{5}{8}$
8. $15\dfrac{7}{24}$
9. $\dfrac{3}{5}$
10. $\dfrac{1}{7}$
11. $2\dfrac{5}{6}$
12. $\dfrac{3}{8}$
13. $1\dfrac{56}{115}$

2.3 ACT Problems:
1. **A**. $\dfrac{2}{3},\dfrac{1}{2},\dfrac{7}{15},\dfrac{2}{7}$ If all else fails, use a calculator to convert them to decimals. $\dfrac{2}{3}=.667$ $\dfrac{1}{2}=.5$ $\dfrac{7}{15}=.4667$ $\dfrac{2}{7}=.2857$

2. J. $8\frac{7}{20}$ The least common denominator is

20. $2\frac{2}{5}=2\frac{8}{20}$ $1\frac{3}{4}=1\frac{15}{20}$ $3\frac{1}{2}=3\frac{10}{20}$

$\frac{7}{10}=\frac{14}{20}$ When you add you get

$6\frac{47}{20}=8\frac{7}{20}$

3. D. $\frac{1}{3}$ 8 pieces are eaten, and 4 pieces are

left out of 12. $\frac{4}{12}=\frac{1}{3}$

4. H. $1\frac{11}{20}$ Half of $3\frac{1}{10}$ is $3\frac{1}{10}\div2$ or

$3\frac{1}{10}\times\frac{1}{2}$ Use an improper fraction.

$\frac{31}{10}\times\frac{1}{2}=\frac{31}{20}=1\frac{11}{20}$

5. A. $\frac{6}{55}$ $\frac{3}{4}\times\frac{8}{15}\div\frac{11}{3}=?$ Flip for division

and reduce.

$\frac{3}{4}\times\frac{8}{15}\times\frac{3}{11}=\frac{3\times8\times3}{4\times15\times11}=\frac{1\times2\times3}{1\times5\times11}=\frac{6}{55}$

Section 2.4 Factors and Prime Numbers
Practice Problems:
1. 1, 13 prime
2. 1, 3, 7, 21 composite
3. 1, 7, 49 composite
4. 1, 5, 11, 55 composite
5. 1, 103 prime
6. 1, 2, 3, 4, 6, 8, 12, 13, 24, 26, 39, 52, 78, 104, 156, 312 composite
7. 1, 3, 5, 15, 37, 111, 185 555 composite
8. 1, 2, 4, 5, 8, 10, 16, 20, 25, 32, 40, 50, 80, 100, 160, 200, 400, 800 composite
9. 1, 2, 4, 5, 8, 10, 20, 25, 40, 50, 100, 125, 200, 250, 500, 1000 composite
10. 1, 3, 677, 2031 composite
11. $12=2^2\cdot3$
12. $30=2\cdot3\cdot5$
13. $39=3\cdot13$
14. $75=3\cdot5^2$
15. $90=2\cdot3^2\cdot5$
16. $98=2\cdot7^2$
17. $121=11^2$
18. $800=2^5\cdot5^2$

2.4 ACT Problems:
1. **D.** 51 $51=3\cdot17$ 51 has factors other than 1 and 51, so it's not prime.
2. **J.** 48 The factors of 35 are 1, 5, 7, and 35. $1+5+7+35=48$
3. **C.** 13 The prime factorization of 624 is $624=2^4\cdot3\cdot13$ The largest prime factor is 13.
4. **G.** 420 $420\div2=210$ $420\div3=140$ $420\div4=105$ $420\div5=84$ $420\div7=60$ 420 is the smallest number that works for all the factors.
5. **C.** 5 The common factors of 28 and 42 are 2, 7, and 14. 2 can divide evenly into 12, so a can't be 2. 7 can divide evenly into 21, so a can't be 7. The only number that works for a is 14. The sum of the digits is $1+4=5$.

Section 2.5 Square Roots
Practice Problems:
1. 5
2. 7
3. 9
4. 12
5. 20
6. $2<\sqrt{5}<3$
7. $3<\sqrt{10}<4$
8. $4<\sqrt{19}<5$
9. $10<\sqrt{111}<11$
10. $11<\sqrt{125}<12$

2.5 ACT Problems:
1. **D.** 1.20 miles $1.2\cdot1.2=1.44$ or $\sqrt{1.44}=1.2$
2. **H.** $\frac{5}{7}$ $\sqrt{\frac{25}{49}}=\frac{\sqrt{25}}{\sqrt{49}}=\frac{5}{7}$
3. **B.** 6 6 rows of 6 cookies each is $6\times6=36$ or $\sqrt{36}=6$
4. **F.** $7<x<8$ $7^2=49$ and $8^2=64$ $49<52<64$ so $\sqrt{49}<\sqrt{52}<\sqrt{64}$

Section 2.6 Exponents and Scientific Notation
Practice Problems:
1. 512
2. 625
3. 10,000,000

4. $\dfrac{1}{11}$

5. $\dfrac{1}{16}$

6. $\dfrac{1}{100,000}$

7. 5

8. 10

9. $\dfrac{1}{2}$

10. 6.2022×10^{1}

11. 3.267×10^{4}

12. 1.124×10^{-6}

13. 682.1

14. 0.0003005

15. 5,599,000

2.6 ACT Problems:

1. **D.** 3×10^{8} m/s Move the decimal 8 times so that it's behind the 3. The number was big, so the exponent is positive.

2. **F.** 5 Remember that an exponent of $\dfrac{1}{2}$ is a square root. $\sqrt{16} < \sqrt{24} < \sqrt{25}$ so $4 < \sqrt{24} < 5$ and 24 is closer to 25 than it is to 16, so $\sqrt{24} \approx 5$

3. **B.** 4 $4^{5} = 1024$

4. **K.** 5×10^{-5} "Per" means "divided by," so 50 parts PER million means $50 \div 1,000,000 = 0.00005$ To convert to scientific notation, move the decimal 5 places so it's behind the 5. The number was smaller than 1, so the exponent is negative.

5. **A.** $\dfrac{8}{9}$ $2^{3} = 8$ and $3^{-2} = \dfrac{1}{3^{2}} = \dfrac{1}{9}$ Multiply.

   $(8)\left(\dfrac{1}{9}\right) = \dfrac{8}{9}$

Section 2.7 Order of Operations
Practice Problems:

1. 73
2. 8
3. 9
4. 44
5. 37
6. 27
7. 14

8. 13
9. 3
10. 0

2.7 ACT Problems:

1. **C.** $(3 \times 4)^{2} - 1$ $3^{2} \times 4 - 1 = 9 \times 4 - 1 = 35$
   $3^{2} \times (4-1) = 9 \times 3 = 27$
   $(3 \times 4)^{2} - 1 = 12^{2} - 1 = 143$
   $(3 \times 4 - 1)^{2} = 11^{2} = 121$ **C** is the largest.

2. **G.** 9 $|3(-5) + 4(6)| = ?$ Work the inside of the absolute value first, starting with multiplication. $|-15 + 24| = ?$ Next, add or subtract. $|9| = ?$ Finally, take the absolute value. $|9| = 9$

3. **A.** -4 $|8-5| - |3-10| = ?$ Work the inside of the absolute values first. $|3| - |-7| = ?$ Then take the absolute values. $3 - 7 = ?$ Finally, subtract. $3 - 7 = -4$

4. **H.** 142 $(3 \times 2^{2})^{2} + (6-10) \div 2 = ?$ Start with the parentheses, and do the exponents inside before the multiplication.
   $(12)^{2} + (-4) \div 2 = ?$ Now do the exponent outside the parentheses. $144 + (-4) \div 2 = ?$ Next is division. $144 + (-2) = ?$ Finally, subtract. $144 + (-2) = 142$

5. **D.** 3 $\dfrac{(-2 + 7 \times 5)}{(5 + 2 \times 3)} = ?$ Start with parentheses, and do the multiplication first. $\dfrac{(-2 + 35)}{(5 + 6)} = ?$ Next add and subtract inside the parentheses. $\dfrac{33}{11} = ?$ Last, divide.
   $\dfrac{33}{11} = 3$

Foundations Review

1. **E.** 40 yards Area = side[2] If the area is 100, $10^{2} = 100$, so the sides are each 10 yards. A square has 4 sides, so the total length around the square is 40 yards.

2. **G.** 12.911_____12.899 The 12's are equal, but in the tenths place, .9 > .8, so 12.911 > 12.899

3. **B.** 243 This problem gives more information than you need. He drives from Nacogdoches to Natchitoches (111 miles) and then from Natchitoches to Natchez (132

miles). $111+132=243$ And, yes, they are real cities.

4. **G**. 0.00006042L 6.042×10^{-5} has a negative exponent, so we need to make the number smaller. Move the decimal 5 places to the left. $6.042\times10^{-5}=0.00006042$

5. **E**. $7\dfrac{9}{16}$ Add $2\dfrac{3}{8}+3\dfrac{1}{4}+1\dfrac{15}{16}$. First, find the common denominator.

$$2\dfrac{3}{8}\cdot\dfrac{2}{2}+3\dfrac{1}{4}\cdot\dfrac{4}{4}+1\dfrac{15}{16}=2\dfrac{6}{16}+3\dfrac{4}{16}+1\dfrac{15}{16}$$

Then add the whole numbers and add the numerators. $2\dfrac{6}{16}+3\dfrac{4}{16}+1\dfrac{15}{16}=6\dfrac{25}{16}$

Convert the improper fraction into a mixed number. $6\dfrac{25}{16}=6+1\dfrac{9}{16}=7\dfrac{9}{16}$

6. **K**. 104 The complete list of factors of 63 is 1, 3, 7, 9, 21, and 63. Sum means add. $1+3+7+9+21+63=104$

7. **A**. 180 Product means multiply. $12(-3)(-5)=(-36)(-5)=180$

8. **H**. 2 $|5(-5)+3(8+1)|=|5(-5)+3(9)|=|-25+27|=|2|=2$

9. **C**. 12 The factors of 48 are 1, 2, 3, 4, 6, 8, 12, 16, 24, and 48. The factors of 36 are 1, 2, 3, 4, 6, 9, 12, 18, and 36. The largest one that they have in common is 12.

10. **H**. 78 $2(9-3)^2+\dfrac{11-(-7)}{3}=2(6)^2+\dfrac{11+7}{3}$

$$=2(36)+\dfrac{18}{3}=72+6=78$$

Chapter 3
Pre-Algebra
14 questions

Section 3.1
Ratios, Proportions, and Cross-Multiplication

Ratio is just another word for fraction. A ratio can be written three ways that all mean the same thing.

$$\frac{1}{2} \quad 1:2 \quad 1 \text{ to } 2$$

A **proportion** is two fractions (two ratios) that are set equal to each other.

$$\frac{1}{2} = \frac{x}{6}$$

To solve a proportion, we **cross-multiply**.

Set the two products equal to each other.

To get x by itself, divide both sides by 2.

Reduce.

$$\frac{1}{2} = \frac{x}{6}$$

$$6 = 2x$$

$$\frac{6}{2} = \frac{2x}{2}$$

$$3 = x$$

Example 1
One gallon of paint can cover 300 square feet. If the walls of Allison's bedroom have an area of 460 square feet. How many gallons of paint should Allison buy?

We can use a proportion to solve this.

Always make sure the units match on your two ratios.

$$\frac{gallons}{sq\ ft} = \frac{gallons}{sq\ ft}$$

Plug in the numbers. Make sure the 1 gallon goes with the 300 sq ft.

$$\frac{1}{300} = \frac{x}{460}$$

Cross-multiply.

$$460 = 300x$$

Divide.

$$\frac{460}{300} = \frac{300x}{300}$$

Reduce.

$$1.5333 = x$$

She can't buy 1.5333 gallons, so round up. She should buy 2 gallons of paint.

Example 2
A remote control car travels at 7 meters per second. If the car travels in a straight line for 2 minutes, what distance will it have covered?

Rate problems are proportions because "per" means "divided by."

Set up a proportion with the units.

We can't plug in 2 minutes, so change it to seconds.

$x = 120$ sec. Now plug into the first proportion.

Cross-multiply.

$$\frac{meters}{seconds} = \frac{meters}{seconds}$$

$$\frac{1\,min}{60\,sec} = \frac{2\,min}{x\,sec}$$

$$\frac{7\,m}{1\,sec} = \frac{x}{120\,sec}$$

$$x = 840 \text{ meters}$$

3.1 Practice Problems: Answers on p. 69
Solve the proportion.

1. $\dfrac{2}{3} = \dfrac{x}{9}$

2. $\dfrac{4}{17} = \dfrac{x}{8}$

3. $\dfrac{5}{13} = \dfrac{6}{x}$

4. 2 to 5 is equal to what to 20?
5. A sign at a grocery store states that oranges are 5 for $3.00. How much would 8 oranges cost?
6. In a litter of puppies, there are 3 girls and 2 boys. What is the ratio of boys to girls? What fraction of the puppies are girls?
7. Just after noon, John stood beside an oak tree and measured his shadow and the shadow of the oak to be 18 inches and 195 inches, respectively. If John is 6 feet tall, how tall is the oak tree?

3.1 ACT Problems: Answers on p. 69

1. A recipe for chocolate chip cookies calls for 2 cups of chocolate chips to make 36 cookies. How many cups of chocolate chips are needed to make 60 cookies?
   A. 3
   B. $3\dfrac{1}{3}$
   C. $3\dfrac{1}{2}$
   D. $3\dfrac{2}{3}$
   E. 4

2. There are 300 swimmers registered for a swim team competition. They are divided into four age groups, as shown in the table below.

Age Group	5–7	8–10	11–13	14+
Number of Swimmers	45	65	110	80

   The judges have 75 prizes to award, and they want to award them proportionally according to the number of swimmers in each age group. How many prizes should be allotted to the 14+ age group?
   F. 8
   G. 14
   H. 18
   J. 20
   K. 25

3. Jessica needs to make punch for a graduation party. The recipe calls for 4 one-liter bottles of pineapple juice to 3 one-liter bottles of ginger ale. If Jessica needs 28 liters of punch, how many one-liter bottles of ginger ale should she buy?

A. 3
B. 7
C. 12
D. 24
E. 28

4. A 22-foot board is cut into two pieces whose lengths have a ratio of 2:3. Which is the length of the longer board?

F. $8\frac{4}{5}$ feet

G. $13\frac{1}{5}$ feet

H. $14\frac{2}{3}$ feet

J. $21\frac{1}{3}$ feet

K. 33 feet

5. A teacher uses about $\frac{2}{3}$ of a box of pencils a year for each student who doesn't bring his or her own school supplies. If there are 35 students in her class, and $\frac{3}{5}$ of them do not bring their own school supplies, how many boxes of pencils will the teacher use this year?

A. 10
B. 14
C. 21
D. 24
E. 35

Section 3.2
Percents

Converting fractions, decimals, and percents

To convert fractions to decimals, just divide on a calculator.
$$\frac{2}{3} = 2 \div 3 = 0.\overline{6}$$

To convert decimals to fractions, use a calculator again.
$$.25 \, f \rightarrow d = \frac{1}{4}$$

On a scientific (non-graphing) calculator there should be a button near the top that says $f \rightarrow d$.
On a graphing calculator, press MATH, and then 1: ▶frac, then ENTER.

To convert decimals to percents, move the decimal 2 places to the right. $0.25 \rightarrow 25\%$
To convert percents to decimals, move the decimal 2 places to the left. $30\% \rightarrow 0.30$

One easy way to remember which way to move the decimal is to think of "percent" as "cents," as in dollars and cents.

$$10\% \rightarrow 10¢ \rightarrow \$0.10 \rightarrow 0.10$$

To convert percents to fractions, put the percent over 100.

$$45\% = \frac{45}{100}$$

To convert fractions to percents, convert to decimal first.

$$\frac{3}{4} = .75 = 75\%$$

Example 1
In a group of 55 freshmen, 30 of them are girls. What percentage of the group are girls?

It's easiest to find the fraction of the group that are girls. 30 out of 55 is $\frac{30}{55}$. Convert it to a decimal.

$30 \div 55 = 0.545454...$ Now move the decimal 2 places. $0.545 = 54.5\%$

Percent word problems
Most percent problems can be set up using "of" and "is." **Of means times** and **is means equals**. Don't forget to change percents into decimals before you type them in your calculator.

Example 2
15% of what number is 6?

Translate the word problem into an equation:

$$\begin{array}{cccc} 15\% & \text{of what number} & \text{is} & 6? \\ .15 & \bullet \quad x & = & 6 \end{array}$$

Solve.

$$.15x = 6$$

Divide by .15

$$\frac{.15x}{.15} = \frac{6}{.15}$$
$$x = 40$$

If the word problem doesn't use of and is, set it up using **part** and **whole**. There are two formulas you can use. Use the one that's most familiar or seems easier to you.

$$percent \times whole = part \qquad \text{Or} \qquad \frac{percent}{100} = \frac{part}{whole}$$

(convert to decimal)   (don't convert to decimal)

Example 3
Mark got 80% on a math test. There were 40 questions on the test. How many questions did Mark answer correctly?

This problem doesn't have "of" and "is," but we can use part and whole. The percent is 80%, the whole is 40, and they asked for the part. You can use either of these formulas:

$$percent \times whole = part$$

convert the percent to decimal

$$.80 \times 40 = x$$

$$32 = x$$

$$\frac{percent}{100} = \frac{part}{whole}$$

don't convert the percent

$$\frac{80}{100} = \frac{x}{40}$$

cross-multiply

$$x = 32$$

Using either formula, we find that Mark answered 32 questions correctly.

Percent of increase or decrease

$$percent\ of\ increase\ or\ decrease = \frac{(subtract\ the\ numbers)}{original\ number}$$

Example 4
In one year, Susan grew from 40" to 43." What was the percent increase in Susan's height?

First, subtract the numbers (big minus small).	$43 - 40 = 3$
Divide by the original. She was 40" before she was 43."	$3 \div 40 = 0.075$
Convert to a percentage. Only move the decimal two places.	$0.075 = 7.5\%$

Percent discount and sales tax
To find a discounted price, multiply the percent off (of) times the price to find the amount off of the price. Then subtract. For example, 25% off $100 is $.25 \times 100 = 25$ then $100 - 25 = \$75$.

Finding sales tax is almost the same as finding a discount, except you add the tax to the price instead of subtracting it. For example, 10% tax on $100 is $.10 \times 100 = 10$ then $100 + 10 = \$110$.

Example 5
A shirt retails for $50 and is on sale for 25% off. If the sales tax is 9%, what is the final price of the shirt?

We'll find the discount first. Find 25% of 50.	$.25 \times 50 = 12.5$
The shirt is $12.50 off. Subtract to find the discounted price.	$50 - 12.5 = 37.5$
The shirt is $37.50 plus tax. Find 9% of 37.5.	$.09 \times 37.5 = 3.375$
Round. The tax is $3.38. Add to find the final price.	$37.5 + 3.38 = 40.88$
The final price paid for the shirt is $40.88.	

3.2 Practice Problems: Answers on p. 69

1. Convert 0.6 to a percent and a fraction.

2. Convert 18% to a decimal and a fraction.

3. Convert $\frac{2}{3}$ to a decimal and a percent.

4. What percent of 80 is 60?

5. 500 is 60 percent of what number?

6. What is 150% of 92?

7. A political action group polled 500 eligible voters before an upcoming election. 350 people were in favor of the group's candidate. What percent were in favor?

8. The speed limit in a neighborhood is 25 mph. In order to issue a ticket, a police officer must record a motorist going 40% above the speed limit. How fast must a motorist be going for an officer to issue a ticket?

9. The price of a couch dropped from $999 to $879. What was the percent of decrease?

10. Before a storm passed through a town, the wind speed was 5 mph. During the storm the wind gusted up to 30 mph. What was the percent of increase in the wind speed before and during the storm?

11. A t-shirt costs $13.99. If the sales tax is 9.9%, what is the final cost of the t-shirt?

12. The regular price of a bicycle is $110. If the store has a 20% off sale, what will be the sale price of the bicycle?

3.2 ACT Problems: Answers on p. 69

1. Between 2007 and 2012, 500 homes were built in Monroe. The graph below shows the number of homes built in each of those years. Approximately what percent of those homes were built in 2011?

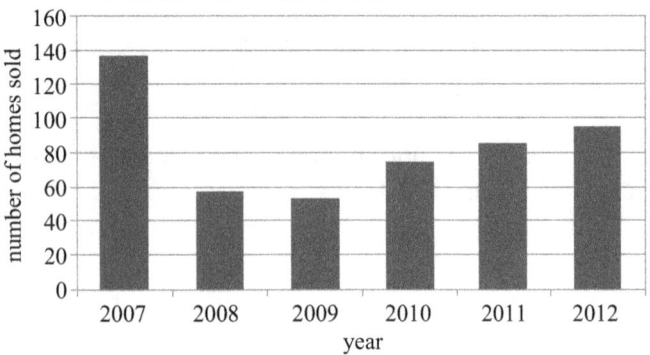

A. 17%
B. 20%
C. 25%
D. 32%
E. 85%

2. What percent of 60 is 6?
F. 6%
G. 10%
H. 12%
J. 15%
K. 25%

3. Ticket sales for the Red County Animal Shelter's annual fund raiser were $250,000 last year. This year, the goal is to raise 30% more than last year from ticket sales. How many dollars does the animal shelter need to raise from ticket sales this year?
A. $280,000
B. $300,000
C. $325,000
D. $350,000
E. $750,000

4. If m is 30% of n, then 130% of n is what percent of m?
F. 130%
G. 231%
H. 330%
J. 433.3%
K. 500%

5. You are waiting to buy a scarf that costs $18.59 when the sales clerk announces that the cash register is broken and she can only accept exact change. You have a 20-dollar bill and a hand full of change. If sales tax is 9%, how much money do you need in coins to buy the scarf?
A. 9¢
B. 21¢
C. 26¢
D. 59¢
E. 67¢

6. A restaurant is trying to cut costs by making its hamburgers smaller. The burgers that used to weigh .25 lbs now weigh .2 lbs. What is the percent decrease in burger weights?
F. .05%
G. .2%
H. 2%
J. 5%
K. 20%

Section 3.3
Mean, Median, and Mode

The **mean** (sometimes called arithmetic mean) of a set of numbers is what we normally think of as the average. Add up all the numbers and then divide by how many numbers there are.

Example 1
Find the mean of these numbers: 12, 15, 21, 14, 15, 11
$$\frac{12+15+21+14+15+11}{6}=\frac{88}{6}=14.67$$

The **median** of a set of numbers is the number in the middle when the numbers are arranged in order. When you drive down an interstate, where's the median? It's in the middle of the road. Don't forget to put the numbers in order!

Example 2
Find the median of these numbers: 12, 15, 21, 14, 15, 11
$$11, 12, (14, 15) 15, 21$$
Average the two in the middle when there is an even number.
$$\frac{14+15}{2}=14.5$$

The **mode** of a set of numbers is the number that repeats the most. It's possible to have no mode if none of the numbers repeats, and it's possible to have more than one mode (bimodal means two modes).

Example 3
Find the mode of these numbers: 12, 15, 21, 14, 15, 11
The mode is 15.

3.3 Practice Problems: Answers on p. 69

Find the mean, median, and mode for these data sets:
1. 1, 2, 3, 4, 5, 6, 7, 8, 9
2. 21, 32, 55, 43, 29, 33, 32, 48
3. 8, 6, 4, 0, 3, 6, 0, 0, 7, 11, 5

4. James received the following test grades: 88, 83, 91, 79, 89. What is his average (arithmetic mean) for the class?
5. A biology class is raising seedlings. The average height of five plants is 3.5 inches. What is the total height of the five plants?
6. On a typical day, a shoe store sells the following sizes: 8, 7, 8, 9, 6, 8, 7, 11, 10, 5, 8, 9, 8, 6, 8. What size shoe should they stock the most?
7. In a school play, a group of students must be arranged by height. The heights are 38 in, 40 in, 44 in, 39 in, 46 in, 36 in, 48 in, 44 in, and 41 in. What height must the student in the middle be?

Use the following information to answer questions 1 – 2.

A real estate company records the number of houses sold over a six month period. The data collected is shown in the graph below.

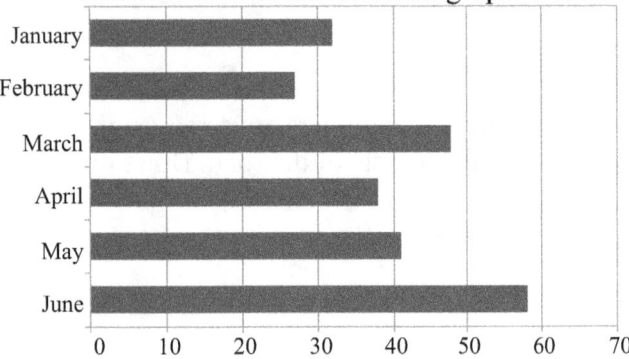

1. What was the average number of homes sold in March, April, and May?
   A. 42
   B. 51
   C. 55
   D. 60
   E. 81

2. The company offered monthly employee bonuses when at least a certain number of houses were sold. If the employees earned the bonuses half of the months shown, what could the goal number of houses sold have been?
   F. 25
   G. 30
   H. 35
   J. 40
   K. 45

3. What is the average of $\frac{5}{16}$ and 0.375?
   A. $\frac{11}{32}$
   B. $\frac{7}{16}$
   C. 0.475
   D. 0.7125
   E. 0.853

4. The average of 8 numbers is 3.2. If each of the numbers is increased by 4.1, what is the new average of the 8 numbers?
   F. 3.2
   G. 3.65
   H. 4.1
   J. 7.3
   K. 8

5. To determine a student's final grade for the year, Mrs. Reitzell discards the student's lowest test score and averages the others. If Noah's test scores in Mrs. Reitzell's class were 75, 86, 91, 83, and 95, what was his final grade?
   A. 83
   B. 86
   C. 89
   D. 92
   E. 95

Section 3.4
Probability

Probability is typically found by making a fraction of $\dfrac{stuff\ you\ want}{total\ stuff\ possible}$ and then reducing or converting to a percent.

Sometimes odds can be written as *probability of what you want* : *probability of what you don't want*. For example, there's a 50:50 chance I'll be free Friday night. 50% chance I'm free : 50% chance I'm not. If the ACT wants you to use this format, they'll tell you. If they don't say, they mean the first way.

Example 1
In a bag, there are 5 green marbles, 3 blue marbles, and 6 yellow marbles. If you reach in the bag and randomly pull out one marble, what is the probability that you draw a yellow marble?

The *stuff you want* is yellow marbles: 6. The *total stuff possible* is the total number of marbles: 14.

$$\frac{stuff\ you\ want}{total\ stuff\ possible}=\frac{6}{14}=\frac{3}{7} \qquad \frac{3}{7}=.4286=42.86\,\%$$

The probability can be correctly written as $\dfrac{3}{7}$ or .4286 or 42.86%. Only one of the three will be an answer choice.

Independent and Dependent Events
Two events are considered **independent** if the outcome of one has no effect on the outcome of the other.

What is the probability that it is raining and I draw a spade from a deck of cards?

Neither of those events can possibly affect the other. They are independent. To find the probability of independent events, just multiply the probabilities together.

Dependent events do have an effect on each other.

What is the probability that the cafeteria serves mystery meat and I have a stomach ache this afternoon?

Do I really need to explain how the first event affects the second? The math can be more complicated on these, but the ACT is unlikely to ask you to calculate a hard one. In general, just multiply the probabilities together.

Example 2
In a bag, there are 5 green marbles, 3 blue marbles, and 6 yellow marbles. If you reach in the bag and randomly pull out one marble, and without replacing it pull out a second, what is the probability that you draw 2 yellow marbles?

These are dependent events. Having one less yellow marble changes the probability for the second draw. The probability for the first draw is $\dfrac{6}{14}$. Assume that you got a yellow marble. Now there are 5 yellow marbles out of a total of 13. The probability for the second draw is $\dfrac{5}{13}$. Multiply.

$$\frac{6}{14}\times\frac{5}{13}=\frac{30}{182} \quad or \quad \frac{15}{91}\ .$$

If you roll a fair six-sided die, what is the probability of rolling:
1. a 5?
2. a 3 or a 6?
3. an even number?

In a standard deck of cards, what is the probability of drawing:
4. a heart?
5. a seven?
6. a black card?
7. a face card?

Note: in a standard deck of cards, there are 52 cards, 4 suits (spades, hearts, diamonds, and clubs), and 13 cards in each suit (ace, king, queen, jack, 10, 9, 8, 7, 6, 5, 4, 3, and 2).

3.4 ACT Problems: Answers on p. 70

1. In Mrs. Robert's desk drawer, there are 9 black pens, 4 blue pens, and 3 red pens. If she reaches into her drawer without looking, what is the probability that she picks up a blue pen?

   A. $\dfrac{1}{2}$

   B. $\dfrac{1}{3}$

   C. $\dfrac{1}{4}$

   D. $\dfrac{1}{8}$

   E. $\dfrac{1}{16}$

2. A jar contains 7 red jelly beans, 5 blue jelly beans, and 8 green jelly beans. If Ollie picks one jelly bean at random, what is the probability that he picks a blue jelly bean?

   F. $\dfrac{1}{4}$

   G. $\dfrac{1}{5}$

   H. $\dfrac{2}{5}$

   J. $\dfrac{1}{20}$

   K. $\dfrac{7}{20}$

3. The circle graph below shows the favorite pizza toppings in Mrs. McHenry's class. If one student is chosen at random, what are the odds that the student's favorite topping is sausage (is favorite : is not favorite)?

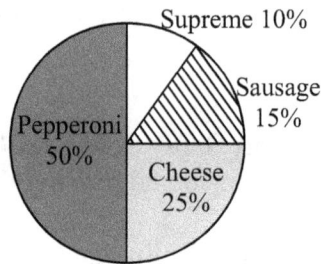

   A. 1:10
   B. 3:21
   C. 3:17
   D. 3:15
   E. 1:4

4. If Todd has already drawn 2 clubs from a standard deck of cards without replacing them, what is the probability that his next card will also be a club?

   F. $\dfrac{1}{4}$

   G. $\dfrac{11}{13}$

   H. $\dfrac{13}{52}$

   J. $\dfrac{13}{50}$

   K. $\dfrac{11}{50}$

5. Anna has a bag of 6 purple marbles, 8 pink marbles, and 10 blue marbles. How many purple marbles does Anna need to add to make the probability of drawing a purple marble $\frac{2}{5}$?

    A. 2
    B. 5
    C. 8
    D. 6
    E. 10

Section 3.5
Organized Counting
AKA Permutations and Combinations

Some problems on the ACT ask you how many ways certain things can be put together. Sometimes the numbers are easy, but other times there are hundreds of possibilities. If you know how to organize the possibilities, you can work these problems in seconds.

On these problems, when in doubt, guess high. It's more than you think it is.

Factorials
A factorial is a shorthand way to say "multiply the number by everything smaller than it."
$$5! = 5 \times 4 \times 3 \times 2 \times 1 = 120$$
ACT likes this because the exclamation mark looks strange. We'll use this later in formulas.

Multiplication Problems
The simplest type of organized counting problem can be solved by just multiplying the choices of one thing by the choices of another thing, and so on.

Example 1
When Juan picks out his school uniform, he can choose from 3 colors of shirt, khaki or navy pants, then sneakers, sandals, or boots. How many different outfits can Juan wear to school?

Just multiply choices of shirt by choices of pants by choices of shoes. $3 \times 2 \times 3 = 18$

Permutations
Use a permutation when you are choosing multiple things from a group and the *order matters* and you *can't* choose the same thing twice. You can either use the formula $\frac{n!}{(n-r)!}$ (preferably on a calculator) or you can just multiply like we did in Example 1.

Example 2
Mr. Wilder's class is choosing a class president, vice president, and secretary. If there are 24 students in the class, how many different ways can the students choose their class officers?

This is a permutation because the order matters (there's a difference between president, vice president, and secretary) and the same student can't be chosen twice.

Method 1: formula
We can use $\dfrac{n!}{(n-r)!}$. n and r are 24 and 3. n is the bigger number or else $n-r$ doesn't make sense.

$$\frac{n!}{(n-r)!} = \frac{24!}{(24-3)!} = \frac{24 \times 23 \times 22 \times \cancel{21} \times \cancel{20} \ldots}{\cancel{21} \times \cancel{20} \ldots} = 24 \times 23 \times 22 = 12144$$

Method 2: multiply
We have 24 choices for president, then 23 choices for v.p. (because the kid picked as president can't be v.p. too) and then 22 choices left for secretary. Multiply.
$24 \times 23 \times 22 = 12144$

Method 3: calculator
Somewhere on your calculator, you should find "nPr." On the TI-30X, look under PRB. On the TI-36X Pro it's on a button above cos. On the TI-83 or TI-84 it's under MATH, right arrow three times to PRB.

Type   24 nPr 3 ENTER. You get 12144.

Combinations
These are like permutations, except the *order doesn't matter*. This formula is more complicated, $\dfrac{n!}{(n-r)!\,r!}$, and you can't just multiply. If the numbers are small, write an organized list. If the numbers are bigger, you'll need to find nCr on your calculator. (It's next to nPr)

Example 3
Mrs. Jenkin's class needs to choose 2 students to sit on student council. 5 of the students want the seats. How many different ways can the class pick 2 students out of the 5?

Method 1: list

We'll call the students A, B, C, D, and E.	AB	BC	CD	DE
Start with A. How many combinations use A?	AC	BD	CE	
4 combinations. Now use B. We can't use BA	AD	BE		
because we already have AB. That's 3 more	AE			
combinations. Keep going. The total is 10.				

Method 2: calculator
Type   5 nCr 2 ENTER. You get 10.

Method 3: formula
If you really want to, you can use this: $\dfrac{n!}{(n-r)!\,r!} = \dfrac{5!}{(5-2)!\,3!} = \dfrac{5 \times 4 \times \cancel{3} \times \cancel{2} \times \cancel{1}}{2 \times 1 \times \cancel{3} \times \cancel{2} \times \cancel{1}} = \dfrac{20}{2} = 10$ ...but why?

1. An ice cream shop has 32 flavors of ice cream, 18 toppings, and the order can be in a sugar cone, waffle cone, or cup. How many different orders with a single scoop and one topping can be placed?

2. How many different license plates can be made using three letters followed by three numbers? The letters and numbers can repeat.

3. How many different 7-digit phone numbers can be made if the first number cannot be a 1 or a 0?

4. How many ways can 7 books be arranged on a shelf?

5. A raffle has three different prizes to be drawn from 82 tickets. How many different ways can three tickets be drawn?

6. You have 6 friends you would like to take out for ice cream, but your car only holds 3 passengers. How many different ways could you choose a group of 3 friends?

3.5 ACT Problems: Answers on p. 70

1. In order to get to their camp site, a group of Boy Scouts must drive on one of three roads, bike along one of five paths, and then hike over one of four trails. How many possible routs are there to take the Boy Scouts to their camp?
   A. 12
   B. 18
   C. 24
   D. 36
   E. 60

2. How many different five digit zip codes can be made if the first digit is not 0? The digits can repeat.
   F. 49
   G. 294
   H. 720
   J. 5,040
   K. 90,000

3. A librarian wants to arrange 3 of the 6 newest books on a display shelf. How many different ways can she arrange the books?
   A. 18
   B. 27
   C. 63
   D. 81
   E. 120

4. 6 students try out for two openings in the school choir. In how many different ways can the openings be filled?
   F. 6
   G. 8
   H. 10
   J. 12
   K. 15

Section 3.6
Solving Linear Equations

Combine Like Terms

3 *apples* + 4 *oranges* + 5 *apples* = 8 *apples* + 4 *oranges* You can't add apples to oranges.

$2x+4y+6+3x-1y-5=5x+3y+1$ You can't add x's to y's or to plain numbers.

Distributing

If you have a number times a parentheses, multiply the number by each term in the parentheses.

$$2(x+4)=2x+8 \qquad\qquad -3(2x-9)=-6x+27$$

Moving Terms

To move a term from one side to the other, do the opposite of it to both sides of the equation.

The opposite of +2 is –2.
$$x+2 = 4$$
$$\underline{-2 \;\; -2}$$
$$x = 2$$

The opposite of ×3 is ÷3.
$$3x=12$$
$$\frac{3x}{3}=\frac{12}{3}$$
$$x = 4$$

Use order of operations backwards.

Example 1

Solve $2x-3+4x+2(x+5)=1-3(6x)+10x+19(2)$

It's long, but just do one piece at a time.
Simplify. Start with the parentheses.

distribute multiply multiply
$$2x-3+4x+2(x+5)=1-3(6x)+10x+19(2)$$

Combine like terms.

$$2x-3+4x+2x+10=1-18x+10x+38$$

Now we start the algebra.

$$8x+7=-8x+39$$

Move the x's to one side...

$$8x+7=-8x+39$$
$$\underline{+8x \qquad\quad +8x}$$

and the numbers to the other.

$$16x+7=39$$
$$\underline{-7 \;\; -7}$$
$$16x=32$$

Finally divide both sides by the number stuck to x.

$$\frac{16x}{16}=\frac{32}{16}$$

Reduce.

$$x=2$$

Example 2

If $4x-3=-2(2+x)$, then $x = ?$

Simplify parentheses.
Don't forget that you're distributing a *negative* 2.
Move the x's.

$$4x-3=-2(2+x)$$
$$4x-3=-4-2x$$
$$\underline{+2x \qquad\qquad +2x}$$
$$6x-3=-4$$

Move the numbers.

$$\underline{+3 \;\; +3}$$
$$6x=-1$$

Divide.

$$\frac{6x}{6}=\frac{-1}{6}$$

Reduce.

$$x=\frac{-1}{6}$$

Solve.

1. $2x+3=5$

2. $13x+7=28$

3. $4-3x=5x-20$

4. $2.5x+1.5=4x-6.0$

5. $2(x+6)=3x$

6. $11x=3(3x+11)$

7. $4(2x-5)+4=2x(3-1)$

8. $11-3(4x+10)=16-(9x+17)$

3.6 ACT Problems: Answers on p. 70-71

1. If $2x-5=13$, then $x=$?
   A. 1.5
   B. 4.0
   C. 8.0
   D. 9.0
   E. 13.0

2. Find an equivalent simplified expression for $5(2x+1)-2(3-x)$.
   F. $12x-1$
   G. $10x-1$
   H. $12x+5$
   J. $8x-1$
   K. $8x+5$

3. If $16-4(3-x)=x-14$, then $x=$?
   A. –8
   B. –6
   C. –2
   D. 2
   E. 4

4. Abby hopes to make $100 from her lemonade stand this weekend. If x represents the number of cups of lemonade sold, she estimates that the profit from her lemonade stand can be found by $(\$.75x)-(\$.25x+\$20)=\$100$. How many cups of lemonade does Abby need to sell this weekend?
   F. 80
   G. 100
   H. 120
   J. 160
   K. 240

5. For what value of p is $x=4$ a solution to the equation $px-8=x+12$?
   A. 3
   B. 4
   C. 5
   D. 6
   E. 7

Section 3.7
Writing Expressions and Equations

Some problems on the ACT give you a word problem and ask for the equation or expression. (An equation has an = sign, but an expression doesn't. Otherwise, they're the same thing.) The word problems are usually long and wordy. You have to pick out the relevant parts. Here are some words and phrases to watch for.

add, sum, more, increased by, in all, total	+
subtract, difference, less, decreased by, how many more (or less)	−
multiply, times, product, of	×
divided by, split, shared, quotient, per	÷
equals, same, is (or any verb)	=

Example 1
Write an equation for this statement: four times the sum of a and b is equal to 8.

We need to multiply and add, but which one first? It says "four times the sum," so we need to add before we multiply by 4. We'll need parentheses for that.
The "equal to 8" is the easy part.
$$4 \times (a+b) = 8$$

Educated Guesses

The good news about these problems is that the answers are long and complicated. Yes, I said *good* news. If you have to guess, *pick the answer that is the most similar to all the other answers*. For example, look at these answer choices:
  A. $5x + 3y$
  B. $5x - 3y$
  C. $3x + 3y$
  D. $5x + 5y$
  E. $15xy$
Answer **E** looks different from all the others. Don't guess that. 3 of the choices have $5x$, so that's probably right. 3 of the choices have $3y$, so that's probably right too. Now we're down to **A** and **B**. The only difference between them is + or −, and there are more +'s, so I would guess **A**.

If you don't know how to work a problem (*any* problem) or if you're running out of time, please use this trick! You'd be surprised how often it works.

Making Stuff Up
With some problems, you can make up easy numbers to plug in and test the answers. Pick easy numbers that you know will work. For example,
   If Ashton is x years old and Grey is y years old, what is their combined age?
Well, if Ashton is 3 and Grey is 5, that's 8 combined.
   $x + y = 3 + 5 = 8$
That works. $x + y$ is the correct answer.

64

Example 2

Whirligigs & Thingamabobs Inc. makes whirligigs and thingamabobs. In analyzing the production schedule for the first quarter of the year, Charlie, the plant manager, wants an expression for the total production cost, in dollars, for making W whirligigs and T thingamabobs. The cost for producing each whirligig is $60, and each thingamabob costs $20 to produce. Which of the following expressions gives the total cost for Whirligigs and Thingamabobs Inc. to produce W whirligigs and T thingamabobs?

    A. $60W - 20T$
    B. $60W + 20T$
    C. $20W + 60T$
    D. $1200(W + T)$
    E. $1200WT$

That problem is way too long. If I cut out the useless parts, I get "...whirligig is $60... thingamabob costs $20... total cost...?" So 60 goes with W and 20 goes with T. To find a total, we add. That's **B**.

If you get overwhelmed by long word problems, and you need to guess, choose the answer that's most similar to all the other answers. **A**, **B**, and **C** are set up alike. **D** and **E** are different, so scratch them out. **A** and **B** agree that the 60 goes with the W, so scratch **C**. Then **B**, **C**, and **D** agree on the + sign. Scratch **A**. We're left with answer **B**, and it's correct.

3.7 Practice Problems: Answers on p. 71

Write an expression or equation for each of the following.

1. 4 times a number, x, less 5
2. 4 times a number, x, less than 5
3. Three more than double a number, n, is three times the number.
4. Double the number of girls, g, is equal to three times the number of boys, b
5. The number of meters (m) per second (t) is the same as the acceleration (a) times the seconds.
6. J. T. is tracking the growth of a plant. Today it is 3 cm high, and it grows .75 cm each day.
7. A salesperson receives a salary of $30,000 plus a commission of 7% of the total sales.
8. A teacher has t treats to give to each of s students. How can she find n, the number of treats that each student gets?
9. Lin has twice as many quarters as dimes. What is the total value of her coins in terms of the number of dimes, d?

3.7 ACT Problems: Answers on p. 71

1. Students are setting up chairs in the gymnasium for the end-of-year program. If they make x rows with $(l + r)$ chairs in each row, how many chairs will they set up in the gymnasium?

    A. $(x \cdot l) + r$
    B. $x + (l + r)$
    C. $(x \cdot l) + (x \cdot r)$
    D. $x \cdot (l \cdot r)$
    E. $x + (l \cdot r)$

2. A physics class does an experiment by rolling a ball along a track. Shamaya records the ball's distance, d, in inches, at set times, t, in seconds, and makes a chart.

t	0	3	6	9	12
d	4	10	16	22	28

Which of the following equations best represents this data?

    F. $d = t + 4$
    G. $d = 2t + 10$
    H. $d = 2t + 4$
    J. $d = 4t + 2$
    K. $d = 10t$

3. When he works steadily, Tom can paint a fence in x hours. If he only works for y hours, what portion of the task remains if y is any number less than x?

A. $(x-y)$

B. $\dfrac{(x-y)}{x}$

C. $\dfrac{(x-y)}{y}$

D. $\dfrac{(x+y)}{x}$

E. $\dfrac{(x-y)}{(x-y)}$

4. For every cent decrease in the price of a snow cone, a snow cone stand sells 10 more snow cones per day. At a price of $.75, the stand sells 100 snow cones a day. Which of the following expressions shows the number of snow cones sold if the price is reduced by p cents?

F. $100(.75+p)$

G. $100-.75$

H. $100+10\,p$

J. $100+.75\,p$

K. $10\,p$

5. A car rental company charges $20 per day and $.20 per mile to rent a compact car. A full size car costs $45 per day and $.25 per mile. Which of the following expressions shows the amount that the cost of a full size car *exceeds* the cost of a compact car when each is rented for d days and driven m miles?

A. $\$45d+\$.25m$

B. $\$25d+\$.05m$

C. $\$25d+\$.50m$

D. $\$25d-\$.05m$

E. $\$20d+\$.20m$

Pre-Algebra Review

1. A number is increased by 55% and the resulting number is decreased by 50%. What percent of the original number is the final number?
   A. 67.5%
   B. 77.5%
   C. 95%
   D. 105%
   E. 110.5%

2. To determine a student's semester grade, Mr. Walker counts the highest test twice and averages the grades. If Santiago has earned grades of 87, 92, 98, 88, and 95, what is his semester grade?
   F. 90
   G. 91
   H. 92
   J. 93
   K. 94

3. Sandy has d dimes and n nickels. Which of the following expressions could be used to find the dollar value of Sandy's coins?
   A. $.10d + .5n$
   B. $.10d + .05n$
   C. $.01d + .05n$
   D. $.15(d + n)$
   E. $.15dn$

4. A basketball team won 80% of its first 20 games. The team lost all of the next 5 games. If their winning percentage was 50% over the last 20 games, what percent of the team's 45 total games did they win?
   F. 26%
   G. 50%
   H. 58%
   J. 75%
   K. 80%

5. A jar contains 5 red marbles, 7 blue marbles, and 9 green marbles. If you pick one marble randomly from the jar, what is the probability that the marble is green?
   A. $\dfrac{3}{7}$
   B. $\dfrac{9}{17}$
   C. $\dfrac{5}{21}$
   D. $\dfrac{1}{3}$
   E. $\dfrac{9}{23}$

6. Pine River Elementary School has set a goal for its fourth graders of reading 500 books in a year. This goal is split among the classes according to the number of students in each class. If there are 60 fourth graders in the school, and 23 of them are in Mrs. Summerton's class, approximately how many books does Mrs. Summerton's class need to read in order to reach their goal?
   F. 115
   G. 192
   H. 221
   J. 250
   K. 323

7. A soccer team has openings for a goalie, a forward, and a midfielder. 6 players are trying out for these positions. How many different ways can the three positions be filled?
   A. 18
   B. 20
   C. 36
   D. 54
   E. 120

8. If $9(2x+5)=6x+21$, then $x = ?$
    F. −3
    G. −2
    H. −1
    J.   1
    K.   2

9. An office supply store is having a sale on graphing calculators. The calculators are marked 24% off. What fraction of the original price is discounted?
    A. $\dfrac{6}{25}$
    B. $\dfrac{1}{24}$
    C. $\dfrac{6}{24}$
    D. $\dfrac{1}{25}$
    E. $\dfrac{25}{6}$

10. Which of the following equations correctly represents this statement?
    5 less than double a number, n, is 8 more than the number.
    F. $5-2n=n+8$
    G. $2n-5=n+8$
    H. $2n-5=8n$
    J. $5n-2=8+n$
    K. $2n+5=8n$

Answers on p. 71-72

Answers and Explanations

Section 3.1 Ratios, Proportions, and Cross-Multiplication
Practice Problems:
1. $x = 6$
2. $x = \dfrac{32}{17}$ or $x = 1.88$
3. $x = \dfrac{78}{5}$ or $x = 15.6$
4. 8
5. $4.80
6. 2:3 (or $\dfrac{2}{3}$ or 2 to 3); $\dfrac{3}{5}$
7. 65 feet

3.1 ACT Problems:
1. **B.** $3\dfrac{1}{3}$ $\dfrac{2}{36} = \dfrac{x}{60}$ $x = 3.3333 = 3\dfrac{1}{3}$
2. **J.** 20 $\dfrac{75}{300} = \dfrac{x}{80}$ $x = 20$
3. **C.** 12 Remember to match up your units. We need $\dfrac{ginger\,ale}{total} = \dfrac{ginger\,ale}{total}$ Total is $4 + 3 = 7$ $\dfrac{3}{7} = \dfrac{x}{28}$ $x = 12$
4. **G.** $13\dfrac{1}{5}$ $\dfrac{long}{total} = \dfrac{long}{total}$ Total is $3 + 2 = 5$ $\dfrac{3}{5} = \dfrac{x}{22}$ $x = 13.2 = 13\dfrac{1}{5}$
5. **B.** 14 Do this one in two parts. $\dfrac{3}{5} = \dfrac{x}{35}$ $x = 21$ students without school supplies. $\dfrac{2}{3} = \dfrac{x}{21}$ $x = 14$ boxes of pencils.

Section 3.2 Percents
Practice Problems:
1. 60%; $\dfrac{3}{5}$
2. .18; $\dfrac{9}{50}$
3. $.\overline{6}$; $66\dfrac{2}{3}$ % or $66.\overline{6}$ %
4. 75%
5. 833.3
6. 138
7. 70%

8. 35 mph
9. 12%
10. 500%
11. $15.38
12. $88

3.2 ACT Problems:
1. **A.** 17% Use "of" and "is" (or "were" in this case). What percent (x) of (•) those homes (500) were (=) built in 2011 (about 85)? $x \cdot 500 = 85$ so $x = .17$ or 17%
2. **G.** 10% Use "of" and "is" again. $x \cdot 60 = 6$ and $x = .1$ which is 10%.
3. **C.** $325,000 The question asks for 30% more, so first find 30% of 250,000. $.30 \cdot 250,000 = 75,000$. $75,000 more than $250,000 is $75,000 + 250,000 = \$325,000$
4. **J.** 433.3% Make up numbers when they give you too many letters. 100 is easy to use with percents. m is 30% of n becomes $30 = .3 \cdot 100$ if $m = 30$ and $n = 100$. Plug these numbers into 130% of n is what percent of m. $1.30 \cdot 100 = x \cdot 30$ Solve for x. $x = 4.333$ or 433.3%
5. **C.** 26¢ First, find 9% of 18.59. $.09 \cdot 18.59 = 1.6731$ Add the tax to the price. $1.67 + 18.59 = 20.26$ You need the $20 bill plus 26¢ in coins.
6. **K.** 20% Subtract the numbers and divide by the original. $\dfrac{0.25 - 0.2}{0.25} = 0.2$ Or 20%

Section 3.3 Mean, Median, and Mode
Practice Problems:
1. mean = 5, median = 5, no mode
2. mean = 36.625, median = 32.5, mode = 32
3. mean = 4.545, median = 5, mode = 0
4. mean = 86
5. total = 17.5
6. mode = 8
7. median = 41 in

3.3 ACT Problems:
1. **A.** 42 March = 48, April = 38, May = 41 $\dfrac{48 + 38 + 41}{3} = 42.3$

2. **J**. 40   Find the median. 27, 32, 38, 41, 48, 58 The median is between 38 and 41.

3. **A**. $\dfrac{11}{32}$ $\dfrac{5/16+0.375}{2}=\dfrac{0.6875}{2}=0.34375$

   $0.34375 \; f{\rightarrow}d = \dfrac{11}{32}$

4. **J**. 7.3   Imagine the numbers are 3.2, 3.2, 3.2, 3.2, 3.2, 3.2, 3.2, and 3.2. Their average is definitely 3.2. If each is increased by 4.1, then the numbers are 7.3, 7.3, 7.3, 7.3, 7.3, 7.3, 7.3, and 7.3. The average is 7.3

5. **C**. 89   Discard the lowest: 75.

   $\dfrac{86+91+83+95}{4}=88.75$

Section 3.4 Probability
Practice Problems:

1. $\dfrac{1}{6}$

2. $\dfrac{2}{6}=\dfrac{1}{3}$

3. $\dfrac{3}{6}=\dfrac{1}{2}$

4. $\dfrac{13}{52}=\dfrac{1}{4}$

5. $\dfrac{4}{52}=\dfrac{1}{13}$

6. $\dfrac{26}{52}=\dfrac{1}{2}$

7. $\dfrac{12}{52}=\dfrac{3}{13}$

3.4 ACT Problems:

1. **C**. $\dfrac{1}{4}$ $\dfrac{blue}{total}=\dfrac{4}{16}=\dfrac{1}{4}$

2. **F**. $\dfrac{1}{4}$ $\dfrac{blue}{total}=\dfrac{5}{20}=\dfrac{1}{4}$

3. **C**. 3:17   *sausage: not sausage* = 15:85 = 3:17

4. **K**. $\dfrac{11}{50}$ If 2 clubs are gone, there are 11 clubs left. If 2 of the total cards are gone, there are 50 left.

5. **D**. 6   Before we add purple marbles, the probability of drawing a purple is $\dfrac{6}{24}$. That does not reduce to $\dfrac{2}{5}$. If we add purple

marbles, that will change the number of purples and the total number. $\dfrac{6+x}{24+x}$. Set that equal to $\dfrac{2}{5}$ and cross-multiply.

$\dfrac{6+x}{24+x}=\dfrac{2}{5}$ $5(6+x)=2(24+x)$

$30+5x=48+2x$ $3x=18$ $x=6$

Section 3.5 Organized Counting
Practice Problems:

1. $32\times18\times3=1{,}728$
2. $26\times26\times26\times10\times10\times10=17{,}576{,}000$
3. $8\times10\times10\times10\times10\times10\times10=8{,}000{,}000$
4. $7!=5{,}040$
5. $82\times81\times80=531{,}360$
6. $6 \; nCr \; 3=20$

3.5 ACT Problems:

1. **E**. 60 $3\,\text{roads}\times5\,\text{bike paths}\times4\,\text{walking trails}=60$ routes

2. **K**. 90,000   There are 9 digits without 0 and 10 digits with 0.
   $9\times10\times10\times10\times10=90{,}000$

3. **E**. 120 $6 \; nPr \; 3=120$ or $6\times5\times4=120$

4. **K**. 15 $6 \; nCr \; 2=15$

Section 3.6 Solving Linear Equations
Practice Problems:

1. $x=1$
2. $x=\dfrac{21}{13}$ or $1\dfrac{8}{13}$
3. $x=3$
4. $x=5$
5. $x=12$
6. $x=\dfrac{33}{2}$ or $16\dfrac{1}{2}$
7. $x=4$
8. $x=-6$

3.6 ACT Problems:

1. **D**. 9.0 $2x-5=13$ Add 5 to both sides. $2x=18$ Divide both sides by 2. $x=9$

2. **F**. $12x-1$ All those big words mean "simplify this": $5(2x+1)-2(3-x)$. Distribute through the parentheses. $10x+5-6+2x$ Combine like terms. $12x-1$ There's no equal sign so we stop here instead of solving.

3. **B.** -6 $16-4(3-x)=x-14$ Distribute over the parentheses. $16-12+4x=x-14$ Combine like terms. $4+4x=x-14$ Subtract x from both sides. $4+3x=-14$ Subtract 4 from both sides. $3x=-18$ Divide both sides by 3. $x=-6$

4. **K.** 240 Pay no attention to the word problem. It says, "Blah, blah, blah, here's an equation. Solve it." We also don't care about dollar signs. $(.75x)-(.25x+20)=100$ Distribute through the parentheses. $.75x-.25x-20=100$ Combine like terms. $.5x-20=100$ Add 20 to both sides. $.5x=120$ Divide both sides by .5 (on a calculator). $x=240$

5. **D.** 6 $px-8=x+12$ We normally can't solve an equation with two variables, but they told us that x is 4. Erase x and write 4. $p4-8=4+12$ Combine like terms. $4p-8=16$ Add 8 to both sides. $4p=24$ Divide both sides by 4. $p=6$

Section 3.7 Writing Linear Expressions and Equations
Practice Problems:
1. $4x-5$
2. $5-4x$
3. $2n+3=3n$
4. $2g=3b$
5. $\dfrac{m}{t}=at$
6. $3+.75t$
7. $30,000+.07s$
8. $n=\dfrac{t}{s}$
9. $.25(2d)+.10d$

3.7 ACT Problems:
1. **C.** $(x\cdot l)+(x\cdot r)$ If you have, for example, 3 rows of 4 chairs, that's $3\times4=12$ chairs. Of course we need to multiply, but $x\cdot(l+r)$ isn't an answer. Distribute the x. You get $(x\cdot l)+(x\cdot r)$.

2. **H.** $d=2t+4$ Just plug in the numbers they gave you, starting with the easiest, $t=0$ and $d=4$. **F** and **H** are the only ones that work. Plug in the next set, $t=3$ and $d=10$. **H** works but **F** doesn't.

3. **B.** $\dfrac{(x-y)}{x}$ Make up some numbers. For example, if it takes 3 hours and he works 2, he has $\dfrac{1}{3}$ left to paint. $\dfrac{(3-2)}{3}=\dfrac{1}{3}$. Answer **B** works.

4. **H.** $100+10p$ This one gives too much information. They didn't ask what the price would be or how much they would make, only how many they would sell. "every cent decrease... sells 10 more." 1 cent sells 10 more, 2 cents sells 20 more, etc. That's $10p$. More than what? More than the 100 they were already selling.

5. **B.** $\$25d+\$.05m$ How much does the price *exceed*? Start with the daily rate. How much does 45 exceed 20? $45-20=25$. We want $25d$, so eliminate **A** and **E**. Now the miles. $.25-.20=.05$. Eliminate **C**. The cost is the daily rate plus the miles, so choose **B**.

Pre-Algebra Review
1. **B.** 77.5% Pick a number. 100 is easy to use with percents. 55% of 100 is 55. Increase 100 by 55 and get 155. Then take 50% of 155. $.5\times155=77.5$. Decrease it, so $155-77.5=77.5$ and $77.5=77.5\%\times100$.

2. **J.** 93 Don't forget to count the highest grade twice. $\dfrac{87+92+98+88+95+98}{6}=93$

3. **B.** $.10d+.05n$ Dimes are worth $.10 and nickels are worth $.05.

4. **H.** 58% First we need to know how many games there were and how many they won. Start with the total games. The first 20 plus 5 plus the last 20 = 45. Now the ones they won. 80% of 20 is 16. ($.80\cdot20=16$) And 50% of 20 is 10. ($.50\cdot20=10$) Add. They won 26. $\dfrac{26}{45}=.57778=58\%$ (or $x\cdot45=26$)

5. **A.** $\dfrac{3}{7}$ $\dfrac{green}{total}=\dfrac{9}{21}=\dfrac{3}{7}$

6. **G.** 192 $\dfrac{part}{whole}=\dfrac{part}{whole}$ $\dfrac{x}{500}=\dfrac{23}{60}$ $x\approx192$

7. **E.** 120 Choices for the first thing times choices for the second times choices for the third. 6 choices for goalie. Once you've

71

picked a goalie, there are 5 players left. 5 choices for forward and 4 for midfielder.

$6 \times 5 \times 4 = 120$ or 6 nPr 3 = 120

8. **G**. –2 $9(2x+5)=6x+21$ Distribute.

$18x+45=6x+21$ Subtract $6x$.

$12x+45=21$ Subtract 45. $12x=-24$

Divide by 12. $x=-2$

9. **A**. $\dfrac{6}{25}$ $24\%=.24=\dfrac{24}{100}=\dfrac{6}{25}$

10. **G**. $2n-5=n+8$ 5 less than double a number, n, is 8 more than the number. "5 less than" means subtract 5. "Double a number" means multiply by 2. The first side has to be $2n-5$. "8 more than" means add 8. The other side has to be $n+8$.

Chapter 4
Elementary Algebra
10 questions

Section 4.1
Evaluating Expressions and Formulas

This section is just plugging numbers into formulas. Sounds easy, right? There are two main ways these questions can stump you. The first problem people have is remembering the formulas. The ACT usually does not give the formula. The second problem is solving for the wrong variable. *Always* take one extra second to check what the question is asking before marking your answer.

A Few Useful Formulas

Triangle Area	$A = \frac{1}{2}bh$	Circle Area	$A = \pi r^2$
Square Area	$A = s^2$	Circumference	$C = \pi d$ or $C = 2\pi r$
Rectangle Area	$A = lw$ or $A = bh$	Prism Volume	$V = (area\ of\ base) \times h$
Parallelogram Area	$A = bh$	Distance, Rate, and Time	$d = rt$
Trapezoid Area	$A = \frac{1}{2}h(b_1 + b_2)$ or $A = bh$ where b = the average base	Simple Interest	$I = PRT$

Example 1
Find the area of a parallelogram with a base of 8 in, a height of 3 in, and a slant height of 5 in.

We can use the formula $A = bh$.

The base is 8 and the height is 3.
The slant height, 5, is not needed to find the area.

Plug in: $A = bh = 8 \times 3 = 24$
The area is 24 in².

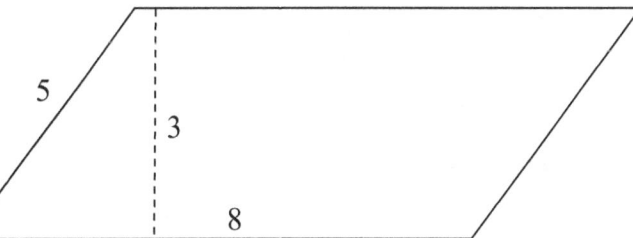

Example 2
Find the area of a triangle with a base of 5 and a height of 10.

We know our formula is $A = \frac{1}{2}bh$, so we plug in our base and height to get $A = \frac{1}{2} \times 5 \times 10 = 25$.

Example 3

What is the interest earned in one year on $1000 invested at an annual interest rate of 5%?

We use the formula $I = PRT$ where I=interest, P=principle (the amount invested), R=rate (as a decimal, not a percent), and T=time (make sure R and T have the same units). For this problem, P=1000, R=.05, and T=1. R is the <u>annual</u> interest rare and T is one <u>year</u>, so the units agree. Plugging in gives us $I = 1000 \times .05 \times 1 = 50$. $50 is earned in one year.

4.1 Practice Problems: Answers on p. 89

1. Find the area of a triangle with base 3 and height 4.
2. Find the area of a triangle with base 9 in and height 13 in.
3. Find the area of a square with side length 4 ft.
4. Find the area of a square with side length 15 cm.
5. Find the area of a rectangle with length 6 and width 2.
6. Find the area of a rectangle with sides 45 yd and 46 yd.
7. Find the area of a parallelogram with base 3 and height 2.
8. Find the area of a parallelogram with base 10, vertical height 4, and slant height 5.
9. Find the area of a trapezoid with bases 6 in and 8 in and height 3 in.
10. Find the area of a circle with radius 5.
11. Find the area of a circle with diameter 6 mm.
12. Find the circumference of a circle with diameter 5 m.
13. Find the circumference of a circle with radius 8 ft.
14. Find the volume of a rectangular prism with length 4 in, width 3 in, and height 5 in.
15. Find the volume of a cube with side length 2 ft.
16. Find the distance traveled in 3 hours at a rate of 50 mph.
17. Find the speed required to travel 120 miles in 2 hours.
18. What is the interest earned on $2,000 in 3 years at an annual interest rate of 8%?
19. How much would you have to invest at 5% annual interest to earn $100 in interest over 2 years?

4.1 ACT Problems: Answers on p. 89

1. What is the area of a circle with a circumference of 14π?
    A. 7π
    B. 28π
    C. 49π
    D. 169π
    E. 196π

2. What is the volume of a cube if the area of one face of the cube is 25 in²?
    F. 5 in³
    G. 10 in³
    H. 50 in³
    J. 100 in³
    K. 125 in³

3. For each day on your newspaper route you receive $5.00 plus a fixed amount for each newspaper you deliver. Currently you earn $13.00 per day when you deliver 80 newspapers. Tomorrow you will deliver 20 additional newspapers. How much will you earn tomorrow?
    A. $5.00
    B. $7.00
    C. $13.20
    D. $15.00
    E. $18.00

4. What is an automobile's average speed, in miles per hour, if it travels 90 miles in 1.5 hours?

F. 35 mph
G. 55 mph
H. 60 mph
J. 75 mph
K. 90 mph

5. Sally defined a new operation, ♥, on pairs of ordered pairs of integers as follows:

$(a,b) ♥ (c,d) = \dfrac{ab+cd}{ac+bd}$. What is the value of $(3,2) ♥ (1,4)$?

A. $\dfrac{3}{5}$

B. $\dfrac{8}{11}$

C. $\dfrac{10}{11}$

D. $\dfrac{11}{10}$

E. $\dfrac{11}{8}$

Section 4.2
Exponents and Radicals

First, remember the vocabulary.

Exponent ↓
Base → $2^3 = 2 \times 2 \times 2$

Index↓
Coefficient→ $2\sqrt[3]{5}$ ←Radicand

Exponents
Be sure you remember these exponent rules:

$$x^0 = 1$$
$$x^1 = x$$
$$x^m \cdot x^n = x^{m+n}$$
$$\frac{x^m}{x^n} = x^{m-n}$$
$$(x^m)^n = x^{mn}$$
$$x^{-m} = \frac{1}{x^m}$$

Example 1
$11^5 \times 11^2 = ?$

Because the bases are the same, we can simply add the exponents to get $11^{5+2} = 11^7$. If the answer choices are simplified, which they usually are, then the easiest way to solve this is to type the original problem in the calculator. $11^\wedge 5 \times 11^\wedge 2$ gives us 19,487,171. Both 11^7 and 19,487,171 are correct. Only one of them will be an answer choice.

Example 2

Simplify $\dfrac{x^3 y^4 x^2}{x^4 y^{-2}}$

Since this question uses variables instead of numbers, we can't use a calculator. First we flip the y with the negative exponent, so we have $\dfrac{x^3 y^4 x^2 y^2}{x^4}$. Then we combine the x's and y's on the top by adding exponents. $\dfrac{x^5 y^6}{x^4}$ Finally we combine the divided x's by subtracting exponents, giving us xy^6 .

Radicals

$$\sqrt[m]{xy} = \sqrt[m]{x} \cdot \sqrt[m]{y}$$
$$\sqrt[n]{\dfrac{x}{y}} = \dfrac{\sqrt[n]{x}}{\sqrt[n]{y}}$$

Radicals can be written as fraction exponents. For example, $5\sqrt[3]{2^4} = 5 \times 2^{\frac{4}{3}}$

When simplifying radicals, reduce the radicand using a factor tree and remove fractions from the denominator. If possible, reduce the index.

Example 3

Simplify $\sqrt{12}$

Using a factor tree, we find that $12 = 2 \cdot 2 \cdot 3$. Since this is a square root, groups of two come out. 12 has two number 2's. Therefore, we get the 2 on the outside and the 3 left inside. $2\sqrt{3}$

```
      12
     /  \
    4    3
   / \
  2   2
12 = (2 · 2) 3
```

Example 4

$\dfrac{6x}{\sqrt{18x^2}} = ?$

First simplify the $\sqrt{18x^2}$ before worrying that the radical is in the denominator.

That gives us $\dfrac{6x}{3x\sqrt{2}}$.

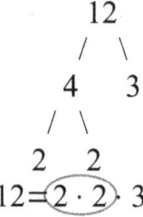

```
      18        x²
     /  \      /  \
    6    3    x    x
   / \
  2   3
18x² = 2 (3·3)(x·x)
```

Next reduce the $6x$ and the $3x$ because they are both outside the radical.

$\dfrac{6x}{3x\sqrt{2}} = \dfrac{2}{\sqrt{2}}$

You cannot reduce the 6 with the 2 because the 2 is inside, but the 6 is outside.

Next, to simplify the denominator, we multiply the top and bottom by $\sqrt{2}$.

$\dfrac{2 \times \sqrt{2}}{\sqrt{2} \times \sqrt{2}} = \dfrac{2\sqrt{2}}{\sqrt{4}} = \dfrac{2\sqrt{2}}{2}$

Finally we reduce the outside 2's leaving $\sqrt{2}$ as our final answer.

$\dfrac{2\sqrt{2}}{2} = \sqrt{2}$

Example 5

Simplify $\sqrt[3]{162}$

```
     162
     / \
    2   81
        / \
       9   9
      / \ / \
     3  3 3  3
```
$162 = 2 \cdot \widehat{3 \cdot 3 \cdot 3} \cdot 3$

Just as before, we start with a factor tree. $162 = 2 \cdot 3 \cdot 3 \cdot 3 \cdot 3$. This is a cube root, so we need groups of three to come outside. We can take three 3's, leaving a 2 and a 3 inside. $3\sqrt[3]{2 \times 3} = 3\sqrt[3]{6}$

4.2 Practice Problems: Answers on p. 89

1. $5^3 \cdot 5^{-1}$

2. $x^3 \cdot x^2$

3. $4^6 + (3^3)^2$

4. $y^2 \div y^2$

5. $12^2 \div 4^2$

6. $9^{1/2} \div 2^{-1}$

7. $\sqrt{36}$

8. $\dfrac{2ab}{\sqrt{144a^3}}$

9. $\sqrt{98}$

10. $\sqrt[3]{56x^5}$

11. $\sqrt[3]{192}$

4.2 ACT Problems: Answers on p. 89

1. $11^{-3} \times 11^3 = ?$
   A. 1
   B. 3
   C. 6
   D. 11
   E. 121

2. $\dfrac{3x}{x^{-2}\sqrt{72}}$
   F. $\dfrac{3x^3}{7}$
   G. $\dfrac{1}{2x^3}$
   H. $\dfrac{2}{3x^{-1}}$
   J. $\dfrac{3x^3\sqrt{6}}{2}$
   K. $\dfrac{x^3\sqrt{2}}{4}$

3. What is the largest value of x for which there exists a real value of y such that $x^2+y^2=225$?
   - A. 15
   - B. 112
   - C. 221
   - D. 225
   - E. 450

4. Which of the following is equal to $\dfrac{4^{10}}{\sqrt{64}}$?
   - F. 130172
   - G. 2^{17}
   - H. 2^{20}
   - J. $8 \cdot 4^{10}$
   - K. 4^{18}

5. All of the following choices are equal to $\dfrac{2^{13} \times 3^2 \sqrt{2}}{18\sqrt{6}}$ EXCEPT
   - A. $(4^3)^2 \times \dfrac{\sqrt{3}}{3}$
   - B. $4^4 \times 4^2 \times \dfrac{\sqrt{3}}{3}$
   - C. $4^5 \times 4 \times \dfrac{\sqrt{3}}{3}$
   - D. $4^8 \div 4^2 \times \dfrac{\sqrt{3}}{3}$
   - E. $4^{12} \div 4^2 \times \dfrac{\sqrt{3}}{3}$

Section 4.3
Operations with Radicals

Addition and Subtraction
In order to combine radicals by addition and subtraction, you must first make the radicals the same. Radicals add and subtract like variables. If the radicals are the same, add or subtract the coefficients. If they are not the same, they cannot combine.
In the examples, I will show you several ways to work these problems. You do not have to understand and use every method. Solve them using the method that makes sense to you.

Example 1
$7\sqrt{5}+2\sqrt{80}=?$
   - A. $-\sqrt{5}$
   - B. $9\sqrt{85}$
   - C. $9\sqrt{5}$
   - D. $15\sqrt{5}$
   - E. $15\sqrt{85}$

–Method 1 – by hand
Always simplify the easier one first. $7\sqrt{5}$ is already in simplest form. Next simplify the other radical using a factor tree. The first number you pull out should be the number left under the radical of the first term (prime or not). The remaining factor (16 in this case) will usually be a perfect square. $\sqrt{80}$ simplifies to $4\sqrt{5}$, but don't forget the 2 that was in front.
$7\sqrt{5}+2\times4\sqrt{5}=7\sqrt{5}+8\sqrt{5}=15\sqrt{5}$

```
      80
     /  \
    5    16
         /  \
        4    4
80 = 4·4·5
```

-Method 2 – calculator comparing decimals

You can simply type the whole problem in the calculator because there are no variables.
$7\sqrt{5}+2\sqrt{80}=33.5410196625$ Now we need to check the answer choices. Our answer is obviously positive, so eliminate **A** first. Then type the remaining choices in your calculator.

- **A.** $-\sqrt{5}$
- **B.** $9\sqrt{85}\ =82.9759$
- **C.** $9\sqrt{5}\ =20.1246$
- **D.** $15\sqrt{5}\ =33.5410$
- **E.** $15\sqrt{85}\ =138.293$

Answer **D** matches our decimal. It must be the answer.

*note: Some calculators will give the answer in radical form. If yours does, just mark the answer!

Example 2

$\sqrt{75}-\sqrt{27}=?$
- **A.** $-2\sqrt{3}$
- **B.** $-\sqrt{3}$
- **C.** $\sqrt{3}$
- **D.** $2\sqrt{3}$
- **E.** $4\sqrt{3}$

-Method 3 – calculator division

Since there are no variables, we can type it in the calculator. $\sqrt{75}-\sqrt{27}=3.46410161514$ The answer is positive, so eliminate **A** and **B**. The remaining answers are all in the form $x\sqrt{3}$. Therefore $3.46410161514=x\sqrt{3}$. If we knew x, we would have our answer. To solve for it, we divide both sides by $\sqrt{3}$.

$\dfrac{x\sqrt{3}}{\sqrt{3}}=\dfrac{3.46410161514}{\sqrt{3}}=2$ The answer must be **D**.

-Method 4 – calculator squaring

Start with the decimal. $\sqrt{75}-\sqrt{27}=3.46410161514$ Eliminate **A** and **B** again. We know a square root is causing our answer to be a decimal. To get rid of a square root, we square it.

$3.46410161514^{2}=12$

We draw the square root back on the 12 and then use a factor tree to get $\sqrt{12}=2\sqrt{3}$.
Again, the answer is **D**.

```
    12
   /  \
  2    6
      /  \
     2    3
12=(2·2)3
```

Multiplication and Division

When multiplying or dividing with radicals, insides go with insides and outsides go with outsides. If both numbers are inside the radical, multiply or divide them. If both numbers are outside the radical, multiply or divide them. After that, simplify with factor trees if needed.

Example 3

$\sqrt{12}\times\sqrt{18}=?$

Both numbers are inside the radicals, so we just multiply them. $\sqrt{12}\times\sqrt{18}=\sqrt{216}$
A factor tree gives us $6\sqrt{6}$.

```
      216
     /  \
    2    108
        /  \
       2    54
           /  \
          9    6
         / \  / \
        3  3 2  3
216=(2·2) 2 (3·3) 3
```

79

Example 4

$$\frac{\sqrt{18}}{\sqrt{12}} = ?$$

Both numbers are inside the radicals, so we reduce them just like any other fraction. $\frac{\sqrt{18}}{\sqrt{12}} = \frac{\sqrt{3}}{\sqrt{2}}$

Remember that a fraction with radicals on the bottom is not simplified. We rationalize the denominator by multiplying the top and bottom by the square root that's on the bottom.

$$\frac{\sqrt{3} \times \sqrt{2}}{\sqrt{2} \times \sqrt{2}} = \frac{\sqrt{6}}{\sqrt{4}} = \frac{\sqrt{6}}{2}$$

Note that the calculator methods shown for addition and subtraction will also work for multiplication and division.

4.3 Practice Problems: Answers on p. 89

1. $3\sqrt{2} + 2\sqrt{8}$

2. $\sqrt{27} + 4\sqrt{48}$

3. $9\sqrt{75} - 11\sqrt{60}$

4. $10\sqrt{432} - 5\sqrt{1323}$

5. $\sqrt[3]{24} \times 2\sqrt[3]{81}$

6. $\sqrt{22} \times \sqrt{14}$

7. $\dfrac{\sqrt{3}}{\sqrt{33}}$

8. $\dfrac{\sqrt{540}}{\sqrt{120}}$

9. $\sqrt{54} + \sqrt{2} \times \sqrt{3}$

10. $\sqrt{8} \times \sqrt{8} - 8 + \sqrt{36}$

4.3 ACT Problems: Answers on p. 89-90

1. $\sqrt{24} - \sqrt{54} = ?$
   A. $-2\sqrt{3}$
   B. $-\sqrt{6}$
   C. $\sqrt{6}$
   D. $2\sqrt{3}$
   E. $\sqrt{30}$

2. $\sqrt{33} \times \sqrt{21} = ?$
   A. $\sqrt{77}$
   B. $2\sqrt{77}$
   C. $3\sqrt{77}$
   D. $4\sqrt{77}$
   E. $5\sqrt{77}$

3. $4\sqrt{80} + 5\sqrt{125} = ?$
   A. $-\sqrt{5}$
   B. $9\sqrt{5}$
   C. $21\sqrt{5}$
   D. $29\sqrt{5}$
   E. $41\sqrt{5}$

4. $\dfrac{\sqrt{17} - \sqrt{68}}{\sqrt{8}} = ?$
   A. $-\sqrt{17}$
   B. $\dfrac{-\sqrt{34}}{4}$
   C. $-\sqrt{34}$
   D. $\sqrt{17}$
   E. $\dfrac{\sqrt{34}}{4}$

5. What is the product of the numerator and

denominator in the simplified form of

$$\frac{\sqrt{72}}{\sqrt{54}} ?$$

    A. $\dfrac{\sqrt{3}}{2}$

    B. $\sqrt{6}$

    C. $2\sqrt{3}$

    D. $6\sqrt{3}$

    E. 12

Section 4.4
Polynomials

Polynomials are just expressions made up of constants and variables combined in various ways.

Terms are constants (coefficients) and variables multiplied or divided by each other.
    Examples: $8, x, -4xy, \dfrac{2z}{3}, x^4y^{-2}z^3$.

Monomials are made of just one term.

Binomials are made of two terms added or subtracted.
    Examples: $2x-3, 4y^2+3x^2, 7a^3b^2c-1$

Trinomials are made of three terms.
    Examples: $x^2+2x+1, y^4-4x^2y^2+3z$

Similar terms are terms with the exact same variable part.
    Examples: $2xy, 3xy,$ and $-5xy$
        $5x^2y$ and $3xy^2$ are NOT like terms.

Adding and Subtracting Polynomials
To add or subtract polynomials, combine like terms. Don't forget to distribute the minus sign when subtracting.

Example 1
$$(2x^3-4x^2+8x-3)-(x^4-3xy^2+6x^2-x^2y)=?$$
No matter which method you use, you should distribute the minus sign.
$$(2x^3-4x^2+8x-3)+(-x^4+3xy^2-6x^2+x^2y)=?$$

-Method 1 – by hand
Group similar terms first. $-x^4+2x^3+(-4x^2-6x^2)+8x+3xy^2+x^2y-3$
Then combine the similar terms. $-x^4+2x^3-10x^2+8x+3xy^2+x^2y-3$

-Method 2 – eliminating answers

Work through these problems one term at a time. The first term is $2x^3$. Are there any more x^3 terms? No, so our answer needs $2x^3$ in it. Stop and look at the answer choices; eliminate any that are missing the $2x^3$ term. Be careful, the ACT isn't nice. They don't always put the terms in order.

The next term is $-4x^2$, so look for more x^2 terms. We have $-6x^2$ also. We combine $-4x^2-6x^2$ to get $-10x^2$. Our answer must have $-10x^2$.

Continue this way until you have eliminated all but one answer. Usually about two terms will be enough.

Multiplying Polynomials
monomial times polynomial

Distribute the monomial and multiply it by each term of the polynomial.

Example 2

Multiply $3x(x^3+4x^2-7x+1)$

Distribute the $3x$ to each term in the parentheses. $3x(x^3+4x^2-7x+1)$
The first term becomes $3x^4$ (stop here and check the answers), then $12x^3$ (check answers again), etc.
The final answer is $3x^4+12x^3-7x^2+3x$

binomial times binomial

Most of you have learned the FOIL method for these. FOIL stands for First, Outside, Inside, Last. You can also think of it as distributing the terms from the first polynomial through the second polynomial one at a time.

Example 3

Multiply $(2x+3)(x-7)$

First: $(2x)(x)=2x^2$
Outside: $(2x)(-7)=-14x$
Inside: $(3)(x)=3x$
Last: $(3)(-7)=-21$

Remember to combine the similar terms in the middle.
$(2x+3)(x-7)=2x^2-14x+3x-21=2x^2-11x-21$

Dividing Polynomials

Using long division on polynomials typically takes too long to be worth it on the ACT. I recommend eliminating answers and making an educated guess. Just like with FOIL, in division the first and last terms are easy; it's the middle that gets messy. Just divide the first terms and the last terms to eliminate answers. Guess from the remaining ones. If you really want a 36 on this test, learn polynomial long division or synthetic division from a textbook.

Example 4

$(x^3 - x^2 - 11x - 10) \div (x + 2) = ?$

We will only work the first and last terms.

First: $\quad (x^3) \div (x) = x^2$

Last: $\quad (-10) \div (2) = -5$

Our answer has x^2 in the front and -5 in the back. Using this method, we can't find the middle term, but we can usually eliminate some answers. The real answer is $x^2 - 3x - 5$ two out of three terms isn't bad compared to the work it would take to find the third term. I would rather you make educated guesses on long problems than run out of time.

4.4 Practice Problems: Answers on p. 90

1. $(3x^4 - 2x^2 + 5) + (x^5 + 7x^4 - 6x^3 - x^2 + 11x - 1)$

2. $(6x^2 y^3 + 8x^2 + 2xy - 2) - (5x^3 y^2 + x^2 - 4xy)$

3. $7x^2 y(3x^2 y^2 - 5x^2 y + 6xy^2 - 9x)$

4. $(x - 3)(3x + 5)$

5. $(x^2 + 12)(x^2 - 12)$

6. $(4x^3 - 8x^2 + 18x - 20) \div (2x)$

4.4 ACT Problems: Answers on p. 90

1. If $M = -4a$ and $N = 2b - a$, what is the value of $M + N$?
   A. $-5a + 2b$
   B. $-5a - 2b$
   C. $-2a - b$
   D. $-2a + b$
   E. $-2a + 2b$

2. What polynomial must be added to $x^2 - 2x + 5$ so that the sum is $3x^2 + 11x$?
   F. $2x^2 + 11x - 5$
   G. $2x^2 + 13x - 5$
   H. $2x^2 - 13x + 5$
   J. $3x^2 + 11x - 5$
   K. $3x^2 - 13x + 5$

3. Which of the following is an equivalent form of $x + x(x + x)$?
   A. $4x^2$
   B. $2x^2$
   C. $x^2 + 2x^2 + x^2$
   D. $2x^2 + x$
   E. $x^2 + 2x$

4. For all real x and m, if $(x - 1)(x + m) = x^2 + kx - m$, then $k = ?$
   F. 1
   G. $-m$
   H. m
   J. $m + 1$
   K. $m - 1$

5. $(2x^3 + 7x^2 - 10x - 24) \div (2x + 3) = ?$
   A. $x^2 + 2x - 8$
   B. $2x^2 + 2x - 8$
   C. $x^2 - 2x + 8$
   D. $x^2 + 8x - 2$
   E. $2x^2 - 8x + 2$

83

Section 4.5
Factoring Polynomials

Common Factors
Do you remember factoring "the easy way," pulling out what the factors have in common?

Example 1
Factor $25x^4y^3 + 30x^2y^7 - 45x^3y^2 + 15x^8y^5$

First look only at the numbers (coefficients). Their greatest common factor is 5. For multiple choice, stop here and eliminate answers that don't have 5 factored out.

Next, see how many x's can come out. Each term has at least 2 x's, so we factor out x^2. Check answers again. Look at the y's. Again, we can factor out two of them. Eliminate answers that don't have y^2. Often, you'll be down to only one answer by now. If so, stop working and mark the answer.

$$5x^2y^2(\;??? \;)$$

If you need the full answer, you need to find what's left from each term when you've factored out the $5x^2y^2$. The first term was $25x^4y^3$. 25 divided by 5 leaves 5, x^4 divided by x^2 leaves x^2 (remember to subtract exponents), and y^3 divided by y^2 leaves y. Divide the other terms the same way.

$$5x^2y^2(5x^2y + 6y^5 - 7x + 3x^6y^3)$$

Factoring Quadratics
This is the factoring with two parentheses, the opposite of FOIL. There are several strategies for this, and if you remember a different method, feel free to use it. It doesn't matter how you get the answer as long as you get the correct answer.

Example 2
Factor $x^2 + 5x + 6$

My strategy is to do the easiest steps first. Step one is to write two empty parentheses.   ( )( )
Next we look at the first terms. To get x^2 we need x and x in the front. $(x \quad)(x \quad)$
For the back, we need two numbers that multiply to give 6 and add up to 5: 2 and 3 $(x \quad 2)(x \quad 3)$
Finally we check the signs. The 6 is positive, so our signs are the same. 5 is positive, $(x + 2)(x + 3)$
so the bigger factor is positive.

Example 3

Factor $6x^2 + x - 12$

Start with the parentheses again. $\qquad$ ( )( )

To get 6 in the front, we need 2 and 3 or 6 and 1. Guess. $\qquad$ $(2x\ \ )(3x\ \ )$

For the 12 in the back we need 3 and 4, 2 and 6, or 1 and 12. Guess. $\qquad$ $(2x\ \ 4)(3x\ \ 3)$

Check the inside and outside terms. Can we get $1x$ from $12x$ and $6x$? No. $\qquad$ $12x$

$\qquad\qquad\qquad\qquad\qquad\qquad\qquad\qquad\qquad\qquad\qquad\qquad\qquad\qquad$ $6x$

Guess again. Switch the 4 and 3. $\qquad$ $(2x\ \ 3)(3x\ \ 4)$

Can we get 1 from 9 and 8? Yes. The 12 was negative, so we subtract. $\qquad$ $9x$

Can we get 1 by subtracting 9 and 8? Yes. The 1 is positive, so we need $\qquad$ $8x$

$+ 9x$ and $- 8x$. $\qquad$ $(2x + 3)(3x - 4)$

Example 4

Factor $(2x + 2 - x + 3)(x + 2) + (x + 5)(2x - 10)$

Once again they are asking a familiar question in an unfamiliar way.

Always check for obvious problems, like the first parentheses. $\qquad$ $(x + 5)(x + 2) + (x + 5)(2x - 10)$

Now treat these as two terms. Do they have anything in common?

Pull out the common factor, $(x + 5)$. $\qquad$ $(x + 5)[(x + 2) + (2x - 10)]$

Add the leftover terms. $\qquad$ $(x + 5)(3x - 8)$

If you get stuck on this one, FOIL the two sides, add like terms, and factor from scratch. It takes longer, but you'll get the same answer.

Sum and Difference of Cubes

$$(x^3 + y^3) = (x + y)(x^2 - xy + y^2) \qquad\qquad (x^3 - y^3) = (x - y)(x^2 + xy + y^2)$$

If you don't know these formulas, let multiple choice help you. Just check the answers.

4.5 Practice Problems: Answers on p. 90

Factor completely.

1. $4x^4 + 12x^2y^2 - 8x^2y^2 + 8x^4$ $\qquad$ 7. $2x^2 + 5x + 2$

2. $21a^3b^2 + 7a^3b^3 - 14a^2b^3 + 35a^2b^2$ $\qquad$ 8. $6x^2 + 13x + 6$

3. $x^2 + 7x + 10$ $\qquad$ 9. $4x^2 + 12x + 9$

4. $x^2 - 8x + 12$ $\qquad$ 10. $5x^2 + 3x - 2$

5. $x^2 + 2x - 8$ $\qquad$ 11. $x^3 + 8$

6. $x^2 - 9$ $\qquad$ 12. $x^3 - 27y^3$

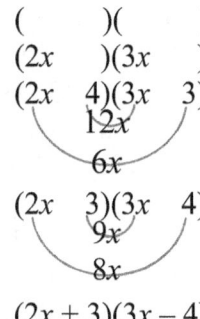

1. Factor $81x^2 - 169y^2$ completely.
   A. $(9x + 13y)(9x - 13y)$
   B. $(9x + 13)(9x - 13)$
   C. $(9 + 13y)(9 - 13y)$
   D. $(9xy + 13)(9xy - 13)$
   E. $(9x - 13y)(9x - 13y)$

2. Factor $4x^2 + 12x - x - 3$ completely.
   F. $(4x + 3)(x - 3)$
   G. $(2x + 3)(2x - 1)$
   H. $(2x + 1)(2x - 3)$
   J. $(4x - 1)(x + 3)$
   K. $(x + 4)(x - 3)$

3. Factor $x^4 - 16$ completely.
   A. $(x^2 + 4)(x^2 - 4)$
   B. $(x + 4)(x - 4)$
   C. $(x^2 + 2)(x^2 - 8)$
   D. $(x + 2)^2(x - 2)^2$
   E. $(x^2 + 4)(x + 2)(x - 2)$

4. Factor $9a^4b^3c + 15a^5b^2c^3 - 3a^2b^2c$ completely.
   F. $9abc(a^3b^2 + 5a^4bc^2 - 3ab)$
   G. $3a^2b^2c(3a^2b + 5a^3c^2 - 1)$
   H. $3abc(3a^3b^2 + 5a^4bc - ab)$
   J. $3a^4b^3c^3(3a + 5b - 3c)$
   K. $9a^4b^3c(6a^5b^2c - 3a^2b^2c)$

5. What is the sum of the factors of $4x^2 - 9$?
   A. $2x + 3$
   B. $2x - 3$
   C. $4x + 6$
   D. $4x$
   E. $4x^2 - 9$

Section 4.6
Solving Quadratic Equations

To solve quadratic equations, we set them equal to zero and factor. That gives us two things multiplied together to make zero. If two numbers multiply to make zero, at least one of the numbers has to be equal to zero. That's why we set each factor equal to zero and solve.

Example 1
Solve $2x^2 + 10x = 0$

We can still factor the easy way first. Pulling out the greatest common factor gives us $2x(x + 5) = 0$
Next we set each factor equal to 0. $2x = 0$ $x + 5 = 0$
Solve. $x = 0$ $x = -5$

Example 2
Solve $x^2 + 5x + 6 = 0$

Factoring gives us $(x + 2)(x + 3)$. See example 2 in Section 4.5.
Set each factor equal to 0. $x + 2 = 0$ $x + 3 = 0$
Solve. $x = -2$ $x = -3$

Solve.

1. $x^2 + 7x + 10 = 0$

5. $2x^2 = -5x - 2$

2. $x^2 - 8x + 12 = 0$

6. $6x^2 + 13x + 6 = 0$

3. $x^2 + 2x = 8$

7. $4x^2 + 12x + 9 = 0$

4. $x^2 = 9$

8. $3x = 2 - 5x^2$

1. What is the solution set to the quadratic equation $x^2 - 144 = 0$?
 A. $\{2, -2\}$
 B. $\{4, -4\}$
 C. $\{9, -9\}$
 D. $\{12, -12\}$
 E. $\{24, -24\}$

2. What is the solution set to the quadratic equation $2x^2 + 5x - 12 = 0$?
 F. $\left\{\dfrac{3}{2}, -4\right\}$
 G. $\left\{\dfrac{2}{3}, -4\right\}$
 H. $\left\{\dfrac{3}{2}, 4\right\}$
 J. $\left\{\dfrac{2}{3}, 4\right\}$
 K. $\{3, 4\}$

3. Which of the following could NOT be the solution set of a quadratic equation?
 A. $\{2\}$
 B. $\{2, -2\}$
 C. $\{1, 2\}$
 D. $\{1, 2, 3\}$
 E. $\{0, 2\}$

4. What is the sum of the solutions of the quadratic equation $x^2 + 5x + 6 = 0$?
 F. -5
 G. -3
 H. 2
 J. 3
 K. 5

5. What is the product of the solutions to the quadratic equation $x^2 + 7x + 10 = 0$?
 A. 2
 B. 5
 C. 7
 D. 10
 E. 17

Elementary Algebra Review

1. What is the sum of the solutions of the equation $x^2 - 3x - 28 = 0$?
   A. -28
   B. -4
   C. 3
   D. 7
   E. 28

2. Which of the following is an equivalent simplified expression for
   $6(2x-4) - 2(4x+7)$
   F. $x + 10$
   G. $2x + 14$
   H. $2x - 38$
   J. $4x + 10$
   K. $4x - 38$

3. Which of the following is a polynomial factor of $4x^2 - 13x - 12$?
   A. $(2x + 3)$
   B. $(2x - 3)$
   C. $(2x + 4)$
   D. $(4x - 3)$
   E. $(4x + 3)$

4. If $a = 3c$ and $b = 6c$, which of the following is true for each nonzero value of c?
   F. $a = 3b$
   G. $a = 2b$
   H. $a = 6b$
   J. $a = \dfrac{1}{3}b$
   K. $a = \dfrac{1}{2}b$

5. If $\sqrt{2x} - 5 = 1$, then $x = $?
   A. 3
   B. 8
   C. 9
   D. 12
   E. 18

6. $\sqrt{x-4}$ is a real number if and only if:
   F. $x \leq -4$
   G. $-4 < x < 0$
   H. $x = 4$
   J. $0 < x < 4$
   K. $x \geq 4$

7. If $r = 2$, $b = 3$, and $y = -4$, what does $(r + b + y)(r - b)$ equal?
   A. -9
   B. -5
   C. -1
   D. 1
   E. 5

8. Which of the following is a factored form of the expression $5x^2 - 13x + 6$?
   F. $(x-3)(5x+2)$
   G. $(x-2)(5x-3)$
   H. $(x-2)(5x+3)$
   J. $(x+2)(5x-3)$
   K. $(x+2)(5x-2)$

9. $4x^5 \cdot 6x^7$ is equivalent to:
   A. $10x^{12}$
   B. $10x^{24}$
   C. $10x^{35}$
   D. $24x^{12}$
   E. $24x^{35}$

10. For a certain fish, the recommended water temperature range is $65° \leq F \leq 80°$. Given the formula $C = 5/9(F-32)$, where C is the temperature in degrees Celsius and F is the temperature in degrees Fahrenheit, what is the corresponding water temperature range in degrees Celsius for the fish?
    F. $0° \leq C \leq 32°$
    G. $18° \leq C \leq 27°$
    H. $27° \leq C \leq 18°$
    J. $27° \leq C \leq 32°$
    K. $32° \leq C \leq 18°$

Answers on p. 91

Answers and Explanations

Section 4.1 Evaluating Expressions and Formulas
Practice Problems
1. 6
2. 58.5 in^2
3. 16 ft^2
4. 225 cm^2
5. 12
6. 2070 yd^2
7. 6
8. 40
9. 21 in^2
10. 25π
11. 9π mm^2
12. 5π m
13. 16π ft
14. 60 in^3
15. 8 ft^3
16. 150 miles
17. 60 mph
18. $480
19. $1000

4.1 ACT Problems
1. **C.** 49π $A=\pi r^2$ and $C=2\pi r$. $14\pi=2\pi r$ so $r=7$. $A=\pi\cdot 7^2=49\pi$
2. **K.** 125 in^3 $V=s^3$ and $A=s^2$. $25=s^2$ so $s=5$. $V=5^3=125$
3. **D.** $15.00 $13=5+80x$. $x=.1$ per newspaper. $S=5+100(.1)=15$
4. **H.** 60 mph $d=r\cdot t$. $90=r(1.5)$ so $r=60$
5. **C.** $\dfrac{10}{11}$ $\dfrac{ab+cd}{ac+bd}=\dfrac{(3)(2)+(1)(4)}{(3)(1)+(2)(4)}=\dfrac{10}{11}$

Section 4.2 Exponents and Radicals
Practice Problems
1. 25
2. x^5
3. 4825
4. 1
5. 9
6. 6
7. 6
8. $\dfrac{b}{6\sqrt{a}}$ or $\dfrac{b\sqrt{a}}{6a}$

9. $7\sqrt{2}$
10. $2x\sqrt[3]{7x^2}$
11. $4\sqrt[3]{3}$

4.2 ACT Problems
1. **A.** 1 $11^{-3}\cdot 11^3=\dfrac{1}{11^3}\cdot 11^3=\dfrac{11^3}{11^3}=1$
2. **K.** $\dfrac{x^3\sqrt{2}}{4}$ $\dfrac{3x}{x^{-2}\sqrt{72}}=\dfrac{3x\cdot x^2}{\sqrt{72}}$
   $\dfrac{3x\cdot x^2}{\sqrt{72}}=\dfrac{3x^3}{6\sqrt{2}}=\dfrac{x^3}{2\sqrt{2}}\times\dfrac{\sqrt{2}}{\sqrt{2}}=\dfrac{x^3\sqrt{2}}{4}$
3. **A.** 15 To get the largest value of x, we need the smallest value of y^2, which is 0. $x^2+0^2=225$ So $x=15$.
4. **G.** 2^{17} $\dfrac{4^{10}}{\sqrt{64}}=\dfrac{1048576}{8}=131072=2^{17}$ Use a calculator to check the answers. Be careful for answers like F that are close but not exact.
5. **E.** $4^{12}\div 4^2\times\dfrac{\sqrt{3}}{3}$ The question asks for the answer that is different. All of the other answer choices simplify to $4^6\times\dfrac{\sqrt{3}}{3}$. **E** simplifies to $4^{10}\times\dfrac{\sqrt{3}}{3}$.

Section 4.3 Operations with Radicals
Practice Problems
1. $7\sqrt{2}$
2. $19\sqrt{3}$
3. $45\sqrt{3}-22\sqrt{15}$
4. $15\sqrt{3}$
5. $12\sqrt[3]{9}$
6. $2\sqrt{77}$
7. $\dfrac{\sqrt{11}}{11}$
8. $\dfrac{3\sqrt{2}}{2}$
9. $4\sqrt{6}$
10. 6

4.3 ACT Problems
1. **B.** $-\sqrt{6}$ $\sqrt{24}-\sqrt{54}=2\sqrt{6}-3\sqrt{6}=-\sqrt{6}$

Or $\sqrt{24}-\sqrt{54}=-2.449489743$
$(-2.449489743)^2=6$

2. **C.** $3\sqrt{77}$ $\sqrt{33}\times\sqrt{21}=\sqrt{693}=3\sqrt{77}$ Or
$\sqrt{33}\times\sqrt{21}=26.324893162$
$\dfrac{26.324893162}{\sqrt{77}}=3$

3. **E.** $41\sqrt{5}$
$4\sqrt{80}+5\sqrt{125}=4(4\sqrt{5})+5(5\sqrt{5})=41\sqrt{5}$
Or $4\sqrt{80}+5\sqrt{125}=91.678787077$
$\dfrac{91.678787077}{\sqrt{5}}=41$

4. **B.** $\dfrac{-\sqrt{34}}{4}$
$\dfrac{\sqrt{17}-\sqrt{68}}{\sqrt{8}}=\dfrac{\sqrt{17}-2\sqrt{17}}{2\sqrt{2}}=\dfrac{-\sqrt{17}}{2\sqrt{2}}=\dfrac{-\sqrt{34}}{4}$
Or $\dfrac{\sqrt{17}-\sqrt{68}}{\sqrt{8}}=-1.457737974$
$-(1.457737974^2)=-2.125=\dfrac{-17}{8}=\dfrac{-34}{16}$
$\dfrac{-\sqrt{34}}{\sqrt{16}}=\dfrac{-\sqrt{34}}{4}$

5. **D.** $6\sqrt{3}$ $\dfrac{\sqrt{72}}{\sqrt{54}}=\dfrac{\sqrt{4}}{\sqrt{3}}=\dfrac{2}{\sqrt{3}}\times\dfrac{\sqrt{3}}{\sqrt{3}}=\dfrac{2\sqrt{3}}{3}$
$2\sqrt{3}\times3=6\sqrt{3}$

Section 4.4 Polynomials
Practice Problems
1. $x^5+10x^4-6x^3-3x^2+11x+4$
2. $6x^2y^3-5x^3y^2+7x^2+6xy-2$
3. $21x^4y^3-35x^4y^2+42x^3y^3-63x^3y$
4. $3x^2-4x-15$
5. x^4-144
6. $2x^2-4x+9-\dfrac{10}{x}$

4.4 ACT Problems
1. **A.** $-5a+2b$
$M+N=(-4a)+(2b-a)=-5a+2b$
2. **G.** $2x^2+13x-5$
$x^2+?=3x^2 \rightarrow 2x^2$. Eliminate **J** and **K**.
$-2x+?=11x \rightarrow +13x$ Choose **G**.
3. **D.** $2x^2+x$
$x+x(x+x)=x+x(2x)=x+2x^2$
4. **K.** $m-1$
$(x-1)(x+m)=x^2+kx-m$
$x^2+mx-x-m=x^2+kx-m$

$mx-x=kx$
$x(m-1)=kx$
$(m-1)=k$

5. **A.** x^2+2x-8
$2x^3\div2x=x^2$ Eliminate **B** and **E**
$-24\div3=-8$ Choose **A**

Section 4.5 Factoring Polynomials
Practice Problems
1. $4x^2(3x^2+y^2)$
2. $7a^2b^2(3a+ab-2b+5)$
3. $(x+5)(x+2)$
4. $(x-6)(x-2)$
5. $(x+4)(x-2)$
6. $(x+3)(x-3)$
7. $(2x+1)(x+2)$
8. $(3x+2)(2x+3)$
9. $(2x+3)^2$
10. $(5x-2)(x+1)$
11. $(x+2)(x^2-2x+4)$
12. $(x-3y)(x^2+3xy+9y^2)$

4.5 ACT Problems
1. **A.** $(9x+13y)(9x-13y)$ **A** is the only answer with the variables in the right places.
2. **J.** $(4x-1)(x+3)$
$(4x-1)(x+3)=4x^2+12x-x-3$
3. **E.** $(x^2+4)(x+2)(x-2)$ Remember to factor (x^2-4)
4. **G.** $3a^2b^2c(3a^2b+5a^3c^2-1)$ The GCF starts with $3a^2$. Choose **G**.
5. **D.** $4x$ $(2x+3)+(2x-3)=4x$

Section 4.6 Solving Quadratic Equations
Practice Problems
1. $x=-2, \ x=-5$
2. $x=6, \ x=2$
3. $x=2, \ x=-4$
4. $x=3, \ x=-3$
5. $x=-\dfrac{1}{2}, \ x=-2$
6. $x=-\dfrac{2}{3}, \ x=-\dfrac{3}{2}$
7. $x=-\dfrac{3}{2}$
8. $x=\dfrac{2}{5}, \ x=-1$

4.6 ACT Problems

1. **D.** $\{12, -12\}$ $x^2 - 144 = (x+12)(x-12) = 0$
   $x + 12 = 0 \quad x - 12 = 0$
   $x = -12 \quad\quad x = 12$

2. **F.** $\left\{\dfrac{3}{2}, -4\right\}$
   $2x^2 + 5x - 12 = (2x-3)(x+4) = 0$
   $2x - 3 = 0 \quad x + 4 = 0$
   $x = \dfrac{3}{2} \quad\quad x = -4$

3. **D.** $\{1, 2, 3\}$ A quadratic equation always gives 2 answers. Sometimes you can get the same answer twice, so there's only one solution, but you can never have 3 solutions.

4. **F.** -5 $x^2 + 5x + 6 = (x+2)(x+3) = 0$
   $x + 2 = 0 \quad x + 3 = 0$
   $x = -2 \quad\quad x = -3 \quad -2 + (-3) = -5$

5. **D.** 10 $x^2 + 7x + 10 = (x+2)(x+5) = 0$
   $x + 2 = 0 \quad x + 5 = 0$
   $x = -2 \quad\quad x = -5 \quad -2(-5) = 10$

Elementary Algebra Review

1. **C.** 3 $x^2 - 3x - 28 = (x+4)(x-7) = 0$
   $x + 4 = 0 \quad x - 7 = 0$
   $x = -4 \quad\quad x = 7 \quad 7 - 4 = 3$

2. **K.** $4x - 38$ $6(2x-4) - 2(4x+7)$
   $12x - 24 - 8x - 14 = 4x - 38$

3. **E.** $(4x + 3)$
   $4x^2 - 13x - 12 = (4x+3)(x-4)$

4. **K.** $a = \dfrac{1}{2}b$ $a = 3c$ and $b = 6c$
   $a = 3c$ and $\dfrac{1}{2}b = 3c$ so $a = \dfrac{1}{2}b$

5. **E.** 18 $\sqrt{2x} - 5 = 1$
   $\sqrt{2x} = 6$
   $2x = 36$
   $x = 18$

6. **K.** $x \geq 4$ A square root is a real number if the expression inside is greater than or equal to zero. $x - 4 \geq 0$ gives $x \geq 4$

7. **C.** -1 $(r + b + y)(r - b) = (2 + 3 - 4)(2 - 3)$
   $= (1)(-1) = -1$

8. **G.** $(x-2)(5x-3)$
   $(x-2)(5x-3) = 5x^2 - 3x - 10x + 6$
   $(x-2)(5x-3) = 5x^2 - 13x + 6$

9. **D.** $24x^{12}$ Multiply the coefficients; add the exponents.

10. **G.** $18° \leq C \leq 27°$ $C = \dfrac{5}{9}(65-32) = 18.333$
    $C = \dfrac{5}{9}(80-32) = 26.667$

91

Chapter 5
Intermediate Algebra
9 questions

Section 5.1
Solving Inequalities

Solving inequalities is exactly like solving regular equations, except that when you multiply or divide both sides by a negative number, you have to turn the sign around.

Example 1
Solve. $x+15\geq18$

$$x+15\geq18$$
Subtract 15. $\dfrac{-15 \quad -15}{}$
$$x\geq3$$

Example 2
Solve. $\dfrac{-a}{5}<6$

Multiply by –5 and flip the sign. $-5\left(\dfrac{-a}{5}\right)>-5(6)$

$$a>-30$$

Example 3
Which is NOT a solution to the inequality $2x+5(x-1)\leq4(x+2)-2x+2$?

    A. 0
    B. 1
    C. 2
    D. 3
    E. 4

	$2x+5x-5\leq4x+8-2x+2$
First simplify and combine like terms on each side separately.	$7x-5\leq2x+10$
Add and subtract from both sides just like an equal sign.	$5x\leq15$
Divide. 5 is positive, so don't change the $\leq$ sign.	$x\leq3$

The question asked for the answer that is NOT a solution, so choose **E. 4**

On questions like this that ask "which is" or "which is NOT a solution," the answer is *usually* (not always) the biggest number or the smallest number. In this case, if 5 or 6 or any other number larger than 3 had been an answer choice, it would have been correct. There can't be more than one choice bigger than 3. The answer has to be the biggest number. If you have to guess and check on this type of problem, try the biggest and smallest numbers first.

1. $x - 3 > 5$

2. $10 - x \leq 19$

3. $3x > 36$

4. $-2x \geq -17$

5. $4(x - 3) > 3x - 8$

6. $\dfrac{3 + x}{4} \geq \dfrac{2x}{5}$

1. What is the solution set for the inequality
   $3x \geq -18$?
   A. $x \leq -6$
   B. $x \geq -6$
   C. $x \leq 6$
   D. $x \geq 6$
   E. $x \geq -54$

2. What is the solution set for the inequality
   $5 - x < 20$?
   F. $x > -25$
   G. $x < -25$
   H. $x > -15$
   J. $x < -15$
   K. $x > -4$

3. Which of the following is not in the
   solution set to the inequality
   $2x + 5 < 4x - 1$?
   A. 3
   B. 4
   C. 5
   D. 6
   E. 7

4. Which of the following is the reciprocal of
   the largest number in the solution set to the
   inequality $3x - 4 \leq -2x + 7$?
   F. $-\dfrac{7}{4}$
   G. $\dfrac{5}{11}$
   H. $\dfrac{11}{5}$
   J. $\dfrac{3}{5}$
   K. $\dfrac{5}{3}$

5. Let p and q be numbers such that $0 < p < q$.
   Which of the following inequalities *must* be
   true for all such p and q?
   A. $p + 1 > q + 1$
   B. $\dfrac{p}{q} > 1$
   C. $\dfrac{1}{q} > \dfrac{1}{p}$
   D. $p^2 > q^2$
   E. $-p > -q$

Section 5.2
Absolute Value Equations and Inequalities

Absolute value makes the number inside positive. If x is inside the absolute value, we don't know if it was positive or negative. This is why we get two answers. For example if $|x|=5$, then $x = 5$ or $x = -5$.

Example 1
$|x+2|=7$

We split this into two equations.
Solve the two equations.

$$x+2=7 \qquad x+2=-7$$
$$x=5 \qquad x=-9$$

Example 2
$|2x-3|<7$

-Method 1

Split the inequalities and change the sign on the negative one.
Solve.

$$2x-3<7 \qquad 2x-3>-7$$
$$x<5 \qquad x>-2$$

Combine the answers. Is it possible to be less than 5 and greater than -2? Yes. Write them together.
If not possible, write separately with "or" between.

$$-2<x<5$$

-Method 2

Change the inequality sign to an equal sign. Split the equations.
Solve.

$$2x-3=7 \qquad 2x-3=-7$$
$$x=5 \qquad x=-2$$

To combine answers, pick a number between 5 and –2.
Plug it in to the original inequality.
It works, so the answer is everything between –2 and 5.
If it doesn't work, write separately with "or" between.

$$|2(0)-3|<7$$
$$3<7$$
$$-2<x<5$$

5.2 Practice Problems: Answers on p. 116

1. $|x-6|=3$

2. $|3x-5|>17$

3. $|8x-5|<9$

4. $|2x+3|\geq 13$

5. $|x-4|\leq 1$

5.2 ACT Problems: Answers on p. 116

1. 14 and –10 are the solutions to which one of the following equations?
   A. $|x-4|=10$
   B. $|x-4|=14$
   C. $|x+4|=10$
   D. $|x-2|=12$
   E. $|x+2|=12$

2. Which of the following makes the inequality $|4x-3|<19$ false?
   F. 0
   G. –1
   H. –2
   J. –3
   K. –4

3. Solve the inequality $|5x+2| \geq 13$
   A. $-3 \leq x \leq 11$
   B. $3 \leq x \leq \dfrac{11}{5}$
   C. $x \leq -3$ or $x \geq 11$
   D. $x \leq -3$ or $x \geq \dfrac{11}{5}$
   E. $-3 \leq x \leq \dfrac{11}{5}$

5. For real numbers m and n, when is the equation $|m-n| = |m+n|$ true?
   A. Always
   B. Only when $m = n$
   C. Only when $m = 0$ and $n = 0$
   D. Only when $m = 0$ or $n = 0$
   E. Never

4. If $|x| > y$, which of the following is the solution statement for y when $x = -3$?
   F. $y = 3$
   G. $y > 3$
   H. $y < 3$
   J. $-3 < y < 3$
   K. $y > 3$ or $y < -3$

Section 5.3
Solving Systems of Equations

A system of equations is a group of two or more equations that must be solved together. If there are two variables, there must be two equations. We usually solve them by substitution or elimination.

Example 1 - Elimination

Solve $4x+3y=11$
        $6x-3y=-6$

When we solve by elimination, we need either the x's or the y's to have the same number with different signs. In this system, we have $3y$ and $-3y$. We can just add the equations.

$$\begin{array}{r} 4x+3y = 11 \\ +\ 6x-3y=-6 \\ \hline 10x\ \ \ \ =5 \\ x=\dfrac{1}{2} \end{array}$$

Now we need to find y. We substitute $x = \dfrac{1}{2}$ into either equation.

$$4\left(\dfrac{1}{2}\right)+3y = 11$$
$$3y = 9$$
$$y = 3$$

Solve.

The solution is written as an ordered pair (x,y). The final answer is $\left(\dfrac{1}{2}, 3\right)$.

Example 2- Substitution

Solve $2x+2y=6$
$y=3x-1$

When we solve by substitution, we solve one of the equations for either x or y. Our second equation is already solved for y, so we just substitute for y in the first equation.

$$2x+2(3x-1)=6$$

Solve.
$$2x+6x-2=6$$
$$8x=8$$
$$x=1$$

Now we substitute $x=1$ in either one of the equations to find y.　　$y=3(1)-1=2$
The final answer is $(1,2)$.

Example 3

Solve $8x-6y=-2$
$-4x+9y+2=7$

It would not be easy to solve for any of the variables, so we'll use elimination. First rewrite the equations so that they're in the same form.

$$8x-6y=-2$$
$$-4x+9y=5$$

Next we need the same number with opposite signs on x or y. 8 and –4 look easier than –6 and 9. Multiply the second equation by 2: $2(-4x+9y=5)$ →
Add.

$$8x-6y=-2$$
$$\underline{-8x+18y=10}$$
$$12y=8$$

Solve.
$$y=\frac{2}{3}$$

Substitute in any of the equations.

$$-4x+9\left(\frac{2}{3}\right)=5$$

Solve.
$$-4x+6=5$$
$$x=\frac{1}{4}$$

The solution is $x=\frac{1}{4}$ and $y=\frac{2}{3}$, or $\left(\frac{1}{4},\frac{2}{3}\right)$

5.3 Practice Problems: Answers on p. 116

1. $3x+5y=8$
    $3x-2y=1$

2. $-2x+y=-6$
    $4x-3y=14$

3. $y=5x-2$
    $y=\frac{1}{2}x+10$

4. $12x-7y=6$
    $4x+3y=2$

5. $8x+y=19$
    $xy=6$

6. Donna has 15 dimes and nickels worth $1.10 . How many of each type of coin does she have?

1. What is the solution to the following system of linear equations?
    $$2x+4y=8$$
    $$3x+4y=9$$
    A. $(1, 1)$
    B. $(1, -1)$
    C. $(1.5, 1)$
    D. $(1.5, 1.5)$
    E. $(1, 1.5)$

2. What is the solution to the following system of linear equations?
    $$6x-3y=9$$
    $$2x+3y=-9$$
    F. $(0, 3)$
    G. $(3, 0)$
    H. $(0, -3)$
    J. $(-3, 0)$
    K. $(-3, 3)$

3. What is the sum of the solutions to the following system of linear equations?
    $$2x+3y=1$$
    $$3x+5y=2$$
    A. -2
    B. -1
    C. 0
    D. 1
    E. 2

4. The sum of x and y is 15. Their difference is 9. What is the value of xy?
    F. 3
    G. 12
    H. 15
    J. 24
    K. 36

5. For what value of a would the following system of equations have an infinite number of solutions?
    $$-2x+3y=-9$$
    $$4x-6y=6a$$
    A. 0
    B. 1
    C. 2
    D. 3
    E. 4

Section 5.4
Matrices

A **matrix** is just a box of numbers. There's nothing magical about them. They can be used to work high-level math problems, but the ACT only tests the basics of matrices.

Adding matrices
Add the numbers that are in the same position. Write the answer in that position. That's all.

$$\begin{bmatrix} 1 & 2 \\ 3 & 4 \end{bmatrix} + \begin{bmatrix} 5 & 6 \\ 7 & 8 \end{bmatrix} = \begin{bmatrix} 6 & 8 \\ 10 & 12 \end{bmatrix}$$

In the top left corners, $1 + 5 = 6$. In the top right, $2 + 6 = 8$. Repeat for the bottom corners.

Example 1

$$\begin{bmatrix} 10 & 14 \\ 8 & -3 \end{bmatrix} + \begin{bmatrix} 11 & -9 \\ 21 & 0 \end{bmatrix} = ?$$

Start with the top left corners. $10 + 11 = 21$ $\begin{bmatrix} 10 & 14 \\ 8 & -3 \end{bmatrix} + \begin{bmatrix} 11 & -9 \\ 21 & 0 \end{bmatrix} = \begin{bmatrix} 21 & \square \\ \square & \square \end{bmatrix}$

Next work the top right corners. $14 - 9 = 5$ $\begin{bmatrix} 10 & 14 \\ 8 & -3 \end{bmatrix} + \begin{bmatrix} 11 & -9 \\ 21 & 0 \end{bmatrix} = \begin{bmatrix} 21 & 5 \\ \square & \square \end{bmatrix}$

Then work the bottom left. $8 + 21 = 29$ $\begin{bmatrix} 10 & 14 \\ 8 & -3 \end{bmatrix} + \begin{bmatrix} 11 & -9 \\ 21 & 0 \end{bmatrix} = \begin{bmatrix} 21 & 5 \\ 29 & \square \end{bmatrix}$

Finally, add the bottom right. $-3 + 0 = -3$ $\begin{bmatrix} 10 & 14 \\ 8 & -3 \end{bmatrix} + \begin{bmatrix} 11 & -9 \\ 21 & 0 \end{bmatrix} = \begin{bmatrix} 21 & 5 \\ 29 & -3 \end{bmatrix}$

On a multiple choice problem, work one corner completely and eliminate answers.

Multiplying a number by a matrix
When you multiply a number by a matrix, the matrix acts like giant parentheses. Distribute the number outside to all the numbers inside.

$$5\begin{bmatrix} 1 & 2 \\ 3 & 4 \end{bmatrix} = \begin{bmatrix} 5 & 10 \\ 15 & 20 \end{bmatrix}$$

In the top left corner, $5(1) = 5$. In the top right, $5(2) = 10$. Repeat for the bottom corners.

Example 2

$$3\begin{bmatrix} 5 & 2 \\ -2 & 11 \end{bmatrix} - \begin{bmatrix} 4 & -12 \\ 2 & 10 \end{bmatrix} = ?$$

Order of operations tells us to multiply first. $3\begin{bmatrix} 5 & 2 \\ -2 & 11 \end{bmatrix} - \begin{bmatrix} 4 & -12 \\ 2 & 10 \end{bmatrix} = ?$

Multiply each number in the first matrix by 3.
Also multiply each number in the second matrix by –1. $\begin{bmatrix} 15 & 6 \\ -6 & 33 \end{bmatrix} + \begin{bmatrix} -4 & 12 \\ -2 & -10 \end{bmatrix} = ?$

Now, add the numbers in the same positions. $\begin{bmatrix} 15 & 6 \\ -6 & 33 \end{bmatrix} + \begin{bmatrix} -4 & 12 \\ -2 & -10 \end{bmatrix} = \begin{bmatrix} 11 & 18 \\ -8 & 23 \end{bmatrix}$

Again, on a multiple choice problem, work one corner completely and eliminate answers.

Multiplying a matrix by a matrix
Multiplying two matrices is more complicated, but ACT rarely tests this. If you have learned this before, it's good to refresh your memory, but if you have never used matrices before, don't worry too much about multiplying them.

To multiply two matrices, the number and length of the *rows* of the first matrix must be the same as the *columns* of the second one. If they're different, it's impossible to multiply them. If they are the same, multiply the *rows* of the first matrix by the *columns* of the second one. Multiply the first number by the first number, then add the second times the second, and so on.

$$\begin{bmatrix} 1 & 2 \\ 3 & 4 \end{bmatrix} \times \begin{bmatrix} 5 & 6 \\ 7 & 8 \end{bmatrix} = ?$$

Start with row 1 times column 1. $[1 \ \ 2] \times \begin{bmatrix} 5 \\ 7 \end{bmatrix} = (1 \times 5) + (2 \times 7) = 5 + 14 = 19$ $\begin{bmatrix} 19 & \square \\ \square & \square \end{bmatrix}$

Next do row 1 times column 2. $[1 \ \ 2] \times \begin{bmatrix} 6 \\ 8 \end{bmatrix} = (1 \times 6) + (2 \times 8) = 6 + 16 = 22$ $\begin{bmatrix} 19 & 22 \\ \square & \square \end{bmatrix}$

Row 2 times column 1. $[3 \ \ 4] \times \begin{bmatrix} 5 \\ 7 \end{bmatrix} = (3 \times 5) + (4 \times 7) = 15 + 28 = 43$ $\begin{bmatrix} 19 & 22 \\ 43 & \square \end{bmatrix}$

Finally, row 2 times column 2. $[3 \ \ 4] \times \begin{bmatrix} 6 \\ 8 \end{bmatrix} = (3 \times 6) + (4 \times 8) = 18 + 32 = 50$ $\begin{bmatrix} 19 & 22 \\ 43 & 50 \end{bmatrix}$

Don't forget to eliminate answers.

Example 3

$$[8 \ \ 12 \ -1] \times \begin{bmatrix} 5 \\ 1 \\ 3 \end{bmatrix} = ?$$

We only have 1 row and 1 column, so our answer will only have 1 row and 1 column. Multiply the first by the first, etc. and add.

$$[8 \ \ 12 \ -1] \times \begin{bmatrix} 5 \\ 1 \\ 3 \end{bmatrix} = (8 \times 5) + (12 \times 1) + (-1 \times 3) = 40 + 12 - 3 = [49]$$

The final answer is 49.

5.4 Practice Problems: Answers on p. 117

1. $\begin{bmatrix} 2 & 3 \\ 6 & 0 \end{bmatrix} + \begin{bmatrix} 3 & 2 \\ 6 & 0 \end{bmatrix} = ?$

2. $\begin{bmatrix} 35 & 18 \\ -61 & 22 \end{bmatrix} + \begin{bmatrix} -9 & 42 \\ 12 & -15 \end{bmatrix} = ?$

3. $4 \begin{bmatrix} -1 & 16 \\ 7 & 8 \end{bmatrix} = ?$

4. $-8 \begin{bmatrix} 12 & 0 \\ -3 & -5 \end{bmatrix} = ?$

5. $5 \begin{bmatrix} -3 & 15 \\ 11 & -2 \end{bmatrix} + \begin{bmatrix} 22 & -9 \\ 17 & 20 \end{bmatrix} = ?$

6. $\frac{1}{2} \begin{bmatrix} -8 & 12 \\ 10 & 4 \end{bmatrix} - \begin{bmatrix} 0 & 2 \\ 1 & 4 \end{bmatrix} = ?$

7. $\begin{bmatrix} 2 & 3 \\ 1 & 2 \end{bmatrix} \times \begin{bmatrix} 1 & 0 \\ 5 & -1 \end{bmatrix} = ?$

8. $[6 \ \ 3 \ \ 4] \times \begin{bmatrix} -1 \\ 4 \\ 8 \end{bmatrix} = ?$

1. $2\begin{bmatrix} 3 & 5 \\ 4 & 6 \end{bmatrix} + \begin{bmatrix} -6 & 11 \\ 10 & -1 \end{bmatrix} = ?$

A. $\begin{bmatrix} 0 & 21 \\ 18 & 11 \end{bmatrix}$

B. $\begin{bmatrix} -3 & 16 \\ 16 & 5 \end{bmatrix}$

C. $\begin{bmatrix} 6 & 10 \\ 8 & 12 \end{bmatrix}$

D. $\begin{bmatrix} -6 & 32 \\ 32 & 10 \end{bmatrix}$

E. $\begin{bmatrix} 0 & -1 \\ -2 & 13 \end{bmatrix}$

2. $3\begin{bmatrix} 10 & -5 \\ 0 & 1 \end{bmatrix} + \begin{bmatrix} 8 & 9 \\ 5 & 2 \end{bmatrix} = ?$

F. $\begin{bmatrix} 18 & 4 \\ 5 & 3 \end{bmatrix}$

G. $\begin{bmatrix} 38 & -6 \\ 5 & 5 \end{bmatrix}$

H. $\begin{bmatrix} 38 & -24 \\ 8 & 5 \end{bmatrix}$

J. $\begin{bmatrix} 30 & -15 \\ 0 & 3 \end{bmatrix}$

K. $\begin{bmatrix} 21 & 7 \\ 8 & 6 \end{bmatrix}$

3. The Twin City Ballet is selling t-shirts, ballet skirts, and bracelets as a fund raiser. The cost and number of items sold are shown below. How much money did the ballet collect after their fund raiser?

Items sold:

$\begin{bmatrix} 25 & 18 & 50 \end{bmatrix}$

T-shirts   Skirts   Bracelets

Item cost:

T-shirts $\begin{bmatrix} \$22 \\ \$12 \\ \$3 \end{bmatrix}$
Skirts
Bracelets

A. $\ 130

B. $\ 846

C. $\ 916

D. $1147

E. $1391

Section 5.5
Exponents and Logs

Exponents

When you multiply or divide numbers or variables with the same bases, you can combine the exponents. Remember these rules:

$$x^m \times x^n = x^{(m+n)}$$
$$(x^m)^n = x^{mn}$$
$$\frac{x^m}{x^n} = x^{(m-n)}$$

You can also solve for the exponents when the bases are the same. For example, if $2^{2x} = 2^{(x+1)}$, we can set the exponents equal to each other because the bases are equal to each other. $2x = x+1 \rightarrow x=1$. When the bases are not equal, we need to use logs.

Logs

Logs are the only way to solve for an exponent.

The two forms for exponential equations are these: $b^x = n$ and $\log_b(n) = x$

For example, $2^x = 3$ can be written as $\log_2(3) = x$.

To solve this on a calculator, we need this formula: $\log_b(n) = \dfrac{\log(n)}{\log(b)}$

So our example is $\log_2(3) = \dfrac{\log(3)}{\log(2)} \approx 1.585$

These are the less common log formulas:
They are not tested often on the ACT. If you have learned
them before, refresh your memory. If not, don't worry
about them.

$$\log(x^a) = a\log(x)$$
$$\log(xy) = \log(x) + \log(y)$$
$$\log\left(\frac{x}{y}\right) = \log(x) - \log(y)$$

Example 1

For all x, $\left(x^{2a+1}\right)^3 = x^9$. $a = ?$

First, we simplify.

$$x^{3(2a+1)} = x^9$$
$$x^{6a+3} = x^9$$

The bases are the same, so set the exponents equal. $6a+3 = 9$

Solve. $a = 1$

Example 2

What is the value of $\log_2(16)$?

There are two ways to work this. If you are working any log or $\log_2(16) = x$

exponential problem by hand, when you get stuck, write it the other way. $2^x = 16$

2 raised to what power equals 16? $2^4 = 16$. $x = 4$

You can also work this on a calculator. $\log_2(16) = \dfrac{\log(16)}{\log(2)} = 4$

Example 3

Condense the following into a single logarithm.

$\log_3(4) + 2\log_3(x) - \log_3(5)$

Start with the 2 multiplied by the second log. $\log_3(4) + 2\log_3(x) - \log_3(5)$

Multiplication turns into an exponent inside the log. $\log_3(4) + \log_3(x^2) - \log_3(5)$

Now addition becomes multiplication and $\log_3(4x^2) - \log_3(5)$

subtraction becomes division inside the log. $\log_3\left(\dfrac{4x^2}{5}\right)$

Simplify.

1. $(a^2 b^3 c^4)^3 (3 a b^2 c)^2$

2. $\dfrac{2 a b (a^5 b^2)}{a^3 b^{-2}}$

3. $2^x \times 4^x$

4. $\log_2(x) + \log_2(3x) - 2\log_2(x)$

5. $2\log(a) - \log(a^2)$

Solve.

6. $2^x = 32$

7. $\log_x(36) = 2$

8. $\log_5 81 - \log_5 3 = \log_5 x$

9. $3^{(3x+1)} = 9^{(x-1)}$

10. $\log_2 10 = x + \log_2 5$

1. Which of the following is equivalent to $(x-3)^0$ whenever $x \neq 3$?
   A. 0
   B. 1
   C. 2
   D. 3
   E. $x - 3$

2. If b is a positive number such that $\log_b(\dfrac{1}{27}) = -3$, then $b = $?
   F. 2
   G. 3
   H. 27
   J. $\dfrac{1}{3}$
   K. $\dfrac{1}{27}$

3. Combine $\log x + 2\log y - \log 3 + \log 2$ into a single log.
   A. $\log(\dfrac{4xy}{3})$
   B. $\log(12xy)$
   C. $\log(\dfrac{2xy^2}{3})$
   D. $\log(\dfrac{x^2 y^2}{3})$
   E. $\log(\dfrac{xy^2}{6})$

4. What is the real value of x in the equation $\log_2 3 + \log_2 4 = \log_2 x$?
   F. 2
   G. 3
   H. 4
   J. 7
   K. 12

5. What is the approximate value of x in the equation $2^x = 9$?
   A. $0 < x < 1$
   B. $1 < x < 2$
   C. $2 < x < 3$
   D. $3 < x < 4$
   E. $4 < x < 5$

Section 5.6
Rational and Radical Expressions

Simplifying Radical Expressions
Remember these radical and exponent rules:

$\sqrt[n]{x}$ means the nth root of x. On a calculator, look for a button or command labeled $\sqrt[x]{}$. For example, $\sqrt[4]{16}$ is entered as 4 $\sqrt[x]{}$ 16.

$\sqrt[a]{x^b} = x^{\frac{b}{a}}$ For example, $8^{\frac{2}{3}}$ can be entered on a calculator 8 ^ (2 / 3).

$x^{(-n)} = \dfrac{1}{x^n}$

$\sqrt{x} \times \sqrt{x} = \sqrt{x^2} = x$

Rational and Undefined Expressions
A rational expression is something that can be written as a fraction (ratio). A rational expression can be reduced if *multiplied* factors can be reduced. For example, $\dfrac{1+1}{1+2}$ does NOT reduce to $\dfrac{1}{2}$. It's $\dfrac{2}{3}$.

$\dfrac{1 \times 2}{2 \times 2}$ does reduce to $\dfrac{1}{2}$.

A rational expression is undefined when the bottom equals zero.

$\dfrac{\sqrt{x}+5}{x}$ is undefined when $x = 0$.

$\dfrac{2x+3}{x-4}$ is undefined when $x = 4$.

$\dfrac{x^2}{5^x}$ is defined for all real numbers. 5^x never equals zero.

The other way for an expression to be undefined is if there is a negative in the square root. $\sqrt{x}$ Is undefined when $x < 0$.

Example 1

Simplify $\dfrac{x^{\frac{1}{4}}}{x^{\frac{-3}{4}} \sqrt{x}}$.

First, flip the variables with negative exponents.

$$\dfrac{x^{\frac{1}{4}} x^{\frac{3}{4}}}{\sqrt{x}}$$

Multiply the x's in the numerator.

$$\dfrac{x^{\frac{1}{4}+\frac{3}{4}}}{\sqrt{x}} = \dfrac{x^1}{\sqrt{x}}$$

Rationalize the denominator (get the root off the bottom).

$$\dfrac{x}{\sqrt{x}} \times \dfrac{\sqrt{x}}{\sqrt{x}} = \dfrac{x\sqrt{x}}{x}$$

Reduce.

$$\sqrt{x}$$

104

Example 2

For which nonnegative value of x is the expression $\dfrac{1}{16-x^2}$ undefined?

We know a fraction is undefined when the denominator equals zero. $\qquad\qquad 16-x^2=0$

Solve. $\qquad\qquad\qquad\qquad\qquad\qquad\qquad\qquad\qquad\qquad\qquad\quad x^2=16$

The question only asked for the positive answer. Don't over complicate it. $\qquad x=4$

Example 3

For all nonzero x and y, $\dfrac{(8x^2y^3)(-5x^2y^2)}{6x^2y^4}=?$

Simplify the top first. Remember when you multiply bases, you add the exponents. $\qquad \dfrac{-40x^4y^5}{6x^2y^4}$

Reduce the top and bottom. When you divide bases, subtract exponents. $\qquad \dfrac{-20x^2y}{3}$

5.6 Practice Problems: Answers on p. 118

Simplify.

1. $x^{\frac{2}{3}}-\sqrt[3]{x^5}\div x^{\frac{1}{3}}+1$

2. $x^2+\sqrt{x^{-1}}\cdot x^{\frac{1}{2}}$

3. $\sqrt[2]{x^2}+\sqrt[3]{x^2}\cdot\sqrt[4]{x^2}\div\sqrt[5]{x^2}$

Find the values of x for which each of the following is defined.

4. $\dfrac{6x^3}{5}$

5. $\dfrac{\sqrt[3]{x-2}}{4x-8}$

6. $\dfrac{2x+3}{x^2+5x+6}$

7. $\dfrac{x\sqrt{2}}{2^{2x}}$

Simplify and state domain restrictions.

8. $\dfrac{3x^2-3y^2}{-x-y}$

9. $\dfrac{3x-x^2}{x^2+3x-18}$

10. $\dfrac{3x^2+14x+8}{x^2+6x+8}$

1. What are the real numbers such that $\dfrac{5x+2\sqrt{x}}{x^2-1}$ is defined?

   A. All real numbers
   B. All real numbers except 1
   C. All real numbers except 0
   D. All real numbers except -1
   E. All nonnegative real numbers except 1

2. If a and b are real and $\sqrt{3\left(\dfrac{a^2}{b}\right)}=1$, then

   what *must* be true of the value of b?

   F. b must be positive
   G. b must be negative
   H. b must equal $\dfrac{1}{3}$
   J. b must equal 3
   K. b may have any value

3. Which of the following is $\dfrac{1}{\sqrt[3]{x}}\cdot\sqrt[3]{x}-\dfrac{\sqrt[3]{x^2}}{x^{\left(\frac{-1}{3}\right)}}$

   in simplest form?
   A. 0
   B. 1
   C. $1-x$
   D. $x^{\left(\frac{1}{3}\right)}-x^{\left(\frac{2}{3}\right)}$
   E. $x^{\left(\frac{1}{3}\right)}-1$

4. Which of the following is $\dfrac{x^2-4x-5}{x^2+2x+1}$ in

   simplest form ($x\neq-1$)?

   F. $\dfrac{-x-5}{x+1}$
   G. $\dfrac{x+5}{x+1}$
   H. $\dfrac{x-5}{x+1}$
   J. $\dfrac{x+2}{x+1}$
   K. $\dfrac{x-2}{x+1}$

5. For all x in the domain of the expression

   $\dfrac{x+1}{x^3-x}$, this expression is equivalent to

   which of the following?
   A. $\dfrac{1}{x^2}-\dfrac{1}{x^3}$
   B. $\dfrac{1}{x^3}-\dfrac{1}{x}$
   C. $\dfrac{1}{x^2-1}$
   D. $\dfrac{1}{x^2-x}$
   E. $\dfrac{1}{x^3}$

Section 5.7
Complex Numbers

An imaginary number is a square root of a negative number.
$$i = \sqrt{-1}$$
Now we can say, for example, $\sqrt{-4} = 2i$.

We can also square i to get -1.
$$i = \sqrt{-1}$$
$$i^2 = -1$$
$$i^3 = i \times i^2 = -i$$
$$i^4 = i^2 \times i^2 = 1$$
$$i^5 = i \times i^4 = i$$
The pattern repeats. Any i with an exponent can be simplified to $i, -1, -i,$ or 1.

A complex number is a number with a real part and an imaginary part. For example, $2 + 3i$ is a complex number. The standard form of a complex number is $a + bi$.

Addition and Subtraction
When working problems with i in them, treat i like a variable. Combine like terms.
Example 1
$(1 + 2i) + (3 - 4i) = ?$
Combine like terms.
$$(1 + 2i) + (3 - 4i) = (1 + 3) + (2i - 4i) = 4 - 2i$$

Example 2
$(-2 + 5i) - (4 + i) = ?$
Distribute the negative and combine like terms.
$$(-2 + 5i) - (4 + i) = -2 + 5i - 4 - i = -6 + 4i$$

Multiplication
Again, treat i as a variable and FOIL. Remember that $i^2 = -1$.

Example 3
$(5 + 6i)(4 - 3i) = ?$
FOIL
$$(5 + 6i)(4 - 3i) = 20 - 15i + 24i - 18i^2 = 20 - 15i + 24i - 18(-1) = 38 + 9i$$

Division
When we divide, we can't leave i in the denominator because i is a square root. To get i out of the denominator, we multiply top and bottom by the conjugate of the denominator ($a + bi$ and $a - bi$ are conjugates). Multiplying conjugates always makes the middle terms cancel, so we only have real numbers and i^2 left on the bottom, and $i^2 = -1$.

Example 4
$\dfrac{1 + 2i}{1 + i} = ?$

$$\frac{1 + 2i}{1 + i} = \frac{1 + 2i}{1 + i} \times \frac{1 - i}{1 - i} = \frac{1 - i + 2i - 2i^2}{1 - i + i - i^2} = \frac{1 + i + 2}{1 + 1} = \frac{3 + i}{2}$$

The final answer can be written as $\dfrac{3 + i}{2}$ or $\dfrac{3}{2} + \dfrac{i}{2}$.

Calculators

Some calculators can work imaginary numbers for you. The TI-36X Pro, the TI-83, and the TI-84, for example, work imaginary numbers very well when the mode is set on $a + bi$. Find out if your calculator works these problems, and if it does, be sure you know how to use it.

5.7 Practice Problems: Answers on p. 118

Simplify

1. $12i^2$

2. $\sqrt{-36}$

3. $2 + 7i - 3 \times 4 - i^2$

4. $(1 + 2i) + (3 - 4i)$

5. $(5 - 6i) - (7 + 8i)$

6. $(2 + i)(3 - i)$

7. $(3 + 5i)(7 - 2i)$

8. $\dfrac{5}{2+i}$

9. $\dfrac{-3-i}{4+3i}$

10. $\dfrac{(.5+4i)(6-2i)}{1-5i}$

5.7 ACT Problems: Answers on p. 119

1. For the complex number i such that $i^2 = -1$, what is the value of $i^4 + 3i^2$?
   A. -4
   B. -2
   C. 0
   D. 2
   E. 4

2. What is $-2(6 - \sqrt{-25})$ in standard form?
   F. $-12 + 10i$
   G. $-12 - 10i$
   H. $-12 - 5i$
   J. -2
   K. $-2i$

3. $(-5 + 2i)(11 - i) = ?$
   A. $-55 + 27i - 2i^2$
   B. $-55 - 27i - 2i^2$
   C. $-53 + 27i$
   D. $-53 - 27i$
   E. $-57 + 27i$

4. For $i^2 = -1$, $(5 + i)^2 = ?$
   F. 24
   G. 26
   H. $24 + 5i$
   J. $24 + 10i$
   K. $26 + 5i$

5. Simplify $\dfrac{9+5i}{2-3i}$.
   A. $\dfrac{3+37i}{5}$
   B. $\dfrac{3+37i}{13}$
   C. $\dfrac{33+37i}{5}$
   D. $\dfrac{33+37i}{13}$
   E. $-\dfrac{3+37i}{5}$

Section 5.8
Solving Quadratic Equations and Inequalities

A quadratic equation is an equation where the variable is squared. When you solve one of these, you should get two answers. You can get the same answer twice, but if you only have one answer, double check that you didn't miss one.

Solving by Factoring
The easiest way to solve a quadratic equation is to set it equal to zero, factor it, and set the factors equal to zero. (See section 4.6)

Example 1
Solve $x^2 + 3x - 10 = 0$.

The equation is already equal to zero, so factor next. $\quad (x+5)(x-2)=0$
Now set each factor equal to zero. $\quad (x+5)=0 \quad (x-2)=0$
Solve. $\quad x=-5 \quad$ or $\quad x=2$

The Quadratic Formula
When a quadratic can't be factored, use the quadratic formula.

First, arrange the equation in standard form. $\quad ax^2+bx+c=0$

Then plug the coefficients in the quadratic equation. $\quad x=\dfrac{-b\pm\sqrt{b^2-4ac}}{2a}$

Example 2
Solve $x^2 + 3x - 2 = 0$.

First, try to factor. There are no numbers that multiply to give 2 and subtract to give 3.
Since you can't factor, use the quadratic formula.
The equation is in standard form. This gives a, b, and c. $\quad a=1, b=3, c=-2$

Plug into the formula. $\quad x=\dfrac{-(3)\pm\sqrt{(3)^2-4(1)(-2)}}{2(1)}$

Simplify. $\quad x=\dfrac{-3\pm\sqrt{9+8}}{2}$

$\qquad\qquad x=\dfrac{-3\pm\sqrt{17}}{2}$

Because our answer has the "plus or minus" sign, it is two answers.

Example 3
Solve $x^2 + 1 = 0$.

Find a, b, and c. There is no middle term, so $b = 0$. $\quad a=1, b=0, c=1$

Plug into the formula. $\quad x=\dfrac{-(0)\pm\sqrt{(0)^2-4(1)(1)}}{2(1)}$

Simplify. $\quad x=\dfrac{\pm\sqrt{-4}}{2}=\dfrac{\pm2i}{2}=\pm i$

Inequalities

The simplest way to work a quadratic inequality is to solve it as if it were an equation, and then figure out the sign. Once you have the two answers, pick a number in between them. Plug that number into the original inequality. If the inequality is true, the solution is everything between the two answers. If the inequality is false, the solution is everything that is NOT between the two answers.

NOTE It is possible for an inequality to have no solution. If one of your answer choices is "no solution," plug in a number that is not between your answers if the in between one didn't work.

Example 4

What is the solution set of $x^2 - x - 2 < 0$?

First, solve it like an equation.

$$x^2 - x - 2 = 0$$
$$(x - 2)(x + 1) = 0$$
$$x = 2 \text{ and } x = -1$$

Choose a number between 2 and –1, for example, 0.
Plug it in to the original problem.

$$(0)^2 - (0) - 2 < 0$$
$$-2 < 0$$

It works, so the answer is everything between 2 and –1.

$$-1 < x < 2$$

5.8 Practice Problems: Answers on p. 119

1. $x^2 - 49 = 0$

2. $x^2 + 3x - 10 > 0$

3. $x^2 - 2x + 5 = 0$

4. $3x^2 \leq 12$

5. $2x^2 + 4x - 17 \geq 0$

6. $2x^2 + 11x + 5 = 0$

7. $x^2 + 2 = 0$

8. $x^2 - 3x > 0$

9. $2x^2 + 9x + 1 = 0$

10. $10{,}000x^2 \geq 1$

110

1. What is the solution set of $x^2 - 3x - 18 < 0$?
   A. $-3 < x < 6$
   B. $x < -3$ or $x > 6$
   C. $2 < x < 9$
   D. $x < 2$ or $x > 9$
   E. $-6 < x < 3$

2. Use the quadratic formula to find the roots of $x^2 + 9 = 0$.
   F. -3 only
   G. -3 and 3
   H. $9 + 3i$ and $9 - 3i$
   J. $-9i$ and $9i$
   K. $-3i$ and $3i$

3. What is the solution set of $x^2 > 19$?
   A. $-\sqrt{19} < x < \sqrt{19}$
   B. $x < -\sqrt{19}$ or $x > \sqrt{19}$
   C. $-9 < x < 10$
   D. $x < -10$ or $x > 9$
   E. $-10 < x < 9$

4. For what nonzero whole number k does the quadratic equation $x^2 + k^2x + 2k = 0$ have exactly 1 real solution for x?
   F. -2
   G. -1
   H. 1
   J. 2
   K. 4

5. The function $f(x)$ is a cubic polynomial that has the value of 0 when x is 0, 4, and -5. If $f(1) = 36$, which of the following is an expression for $f(x)$?
   A. $x^3 - 25$
   B. $-2x(x - 4)(x + 5)$
   C. $x(x - 4)(x + 5)$
   D. $2x(x + 4)(x - 5)$
   E. $2x(x - 4)(x + 5)$

Section 5.9
Sequences and Series

A sequence is just a list of numbers that follows a pattern. A series is a sequence that is added. For example, 2, 4, 6 is a sequence that follows a + 2 pattern. 2 + 4 + 6 is a series.

Arithmetic Sequences
An arithmetic sequence (or series) follows a pattern of adding a certain number to the previous term. For example, the sequence 6, 9, 12 adds 3 to each term. 3 is called the common difference because subtracting any two consecutive terms gives a difference of 3.

Example 1
Find the next term in the sequence 15, 13, 11, 9, ___

First, we need to find the common difference (or find the pattern). How do you get from 15 to 13? You subtract 2. Does that work to get from 13 to 11? 11 to 9? Yes, we have found the pattern. To get from 9 to the next term, we subtract 2 again. The next term is 7.

Geometric Sequences

A geometric sequence (or series) follows a pattern of multiplying a certain number times the previous term. For example, the sequence 2, 4, 8, 16 multiplies each term by 2. 2 is called the common ratio because dividing any term by the one before it gives a ratio of 2.

Example 2

Find the next term in the sequence 1, –3, 9, –27, __

First, find the pattern. How do you get from 1 to –3? You could subtract 4. Does that work to get from –3 to 9? No, that's not the pattern. How else can you get from 1 to –3? You could multiply by –3. Does that work to get from –3 to 9? Yes, –3 is the common ratio. Multiply –27 times –3 to get the next term, 81.

Finding the Sum of a Series

There are formulas for sequences and series, but the ACT usually uses small numbers that are easy enough to figure out without memorizing a formula. If you can, just add up all the numbers on a calculator.

To add up a longer arithmetic series, find the average of the first and last terms. That is also the average of all the terms. Now treat the series like every term is the average.

Example 3

An arithmetic series' first term is 2, and 50th term is 100. Find the sum of the first 50 terms.

First find the average term: $\dfrac{2+100}{2}=51$. Treat the series as if every term equaled 51. 51 + 51 + 51... 50 times is the same as $51\times50=2550$. The sum of the whole series is 2550.

5.9 Practice Problems: Answers on p. 119

Write the next three terms of the following sequences.

1. 4, 6, 8, 10, …

2. 100, 85, 70, 55, …

3. 8, 4, 2, …

4. $\dfrac{1}{3}$, –1, 3, –9, …

5. 1, 1, 2, 3, 5, 8, …

1. The first term in a geometric sequence is 4, and the common ratio is 3. Which of the following shows the first three terms in the sequence?
   A. 4, 12, 36
   B. 4, 6, 8
   C. 3, 4, 5
   D. 4, 3, 2
   E. 12, 36, 108

2. One day Sam remembers 3 vocabulary words. The second day he remembers 5 vocabulary words, and the third day he remembers 7 vocabulary words. If this pattern continues, how many words will Sam remember on the tenth day?
   F. 10
   G. 19
   H. 21
   J. 23
   K. 30

3. Which of the following choices shows the sixth term in the sequence below?
   5, 30, 180, ...
   A. 130
   B. 630
   C. 6,480
   D. 15,625
   E. 38,880

4. The pattern shown below in an abbreviated form is composed of squares that are arranged horizontally and surrounded by 4 hexagons. All the squares are congruent, and all the hexagons are congruent. How many of these congruent hexagons will there be if the pattern is repeated until there are 20 squares?

   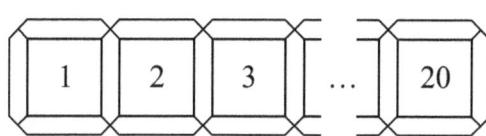

   F. 44
   G. 61
   H. 70
   J. 79
   K. 80

5. Mary runs every week. The table shows the total amount of time Mary runs in four consecutive weeks. If this pattern continues, how many minutes will Mary spend running in the 5th week?

Week 1	75 minutes
Week 2	86 minutes
Week 3	97 minutes
Week 4	108 minutes

   A. 119 minutes
   B. 122 minutes
   C. 125 minutes
   D. 175 minutes
   E. 216 minutes

Intermediate Algebra Review

1. Which of the following identifies exactly those values of x that satisfy
$6x - 3(3x + 2) \leq 5 - 5(3 - x)$?

   A. All real numbers

   B. $x \geq \dfrac{1}{2}$

   C. $x \leq \dfrac{1}{2}$

   D. $x \geq 2$

   E. No real numbers

2. The recursive formula for a sequence is given below, where a_n is the value of the nth term.
$$a_1 = 12$$
$$a_n = a_{n-1} + 6$$
Which of the following equations is an explicit formula for this sequence?

   F. $a_n = -6n + 12$
   G. $a_n = 6n + 6$
   H. $a_n = 6n + 12$
   J. $a_n = 12n - 6$
   K. $a_n = 12n + 6$

3. For all $x > 2$, $\dfrac{2x - x^2}{x^2 + 4x - 12} = ?$

   A. $\dfrac{-x}{x+6}$

   B. $\dfrac{x}{x-6}$

   C. $\dfrac{1}{x+6}$

   D. $\dfrac{-1}{12}$

   E. $\dfrac{1}{12}$

4. What are the (x,y) coordinates of the intersection of the lines determined by the equations $3x + 4y = 22$ and $x + 2y = 12$?

   F. The lines don't intersect.
   G. $(-2, -7)$
   H. $(-2, 7)$
   J. $(2, 7)$
   K. There are infinitely many points in the intersection.

5. Which of the following is not a solution to the inequality $x^2 - 5x + 4 \leq 0$?

   A. 0
   B. 1
   C. 2
   D. 3
   E. 4

6. If, for all x, $(x^{2a-1})^4 = x^{12}$, then $a = ?$

   F. 0.5
   G. 1
   H. 1.5
   J. 2
   K. 2.5

7. Sometimes it is more convenient to work with fractions that have no square roots in the denominator. Which of the following is an equivalent form for $\dfrac{6x}{\sqrt{6x} - y}$?

   A. $\dfrac{-\sqrt{6x} - y}{y}$

   B. $\dfrac{\sqrt{6x} - y}{6x}$

   C. $\dfrac{6x\sqrt{6x} - y}{36x^2 - y^2}$

   D. $\dfrac{6x\sqrt{6x} + y}{6x - y}$

   E. $\dfrac{6x\sqrt{6x} - y}{6x - y}$

8. $4\begin{bmatrix} 5 & 7 \\ 6 & 8 \end{bmatrix} + \begin{bmatrix} -1 & 3 \\ 2 & -4 \end{bmatrix} = ?$

    F. $\begin{bmatrix} 4 & 10 \\ 8 & 4 \end{bmatrix}$

    G. $\begin{bmatrix} 20 & 28 \\ 24 & 32 \end{bmatrix}$

    H. $\begin{bmatrix} 19 & 31 \\ 20 & 34 \end{bmatrix}$

    J. $\begin{bmatrix} 16 & 40 \\ 32 & 16 \end{bmatrix}$

    K. $\begin{bmatrix} 19 & 31 \\ 26 & 28 \end{bmatrix}$

9. For $i^2 = -1$, $(5 + 2i)^2 = ?$
    A. $25 + 20i + 4i^2$
    B. $25 + 20i - 4i^2$
    C. $21 + 20i$
    D. $25 + 20i$
    E. $21 + 16i$

Answers on p. 120

10. For all pairs of real numbers x and y, which of the following expressions are equal?
    I. $|x+y|$
    II. $|x|+|y|$
    III. $|-x-y|$
    F. I and II only
    G. I and III only
    H. II and III only
    J. I, II, and III
    K. No pair of expressions will always be equal.

Answers and Explanations

Section 5.1 Solving Inequalities
Practice Problems:
1. $x > 8$
2. $x \geq -9$
3. $x > 12$
4. $x \leq \dfrac{17}{2}$
5. $x > 4$
6. $x \leq 5$

5.1 ACT Problems:
1. **B.** $x \geq -6$ Divide both sides by 3; don't change the sign.
2. **H.** $x > -15$
$$5 - x < 20$$
$$-x < 15$$
$$x > -15$$
3. **A.** 3   Solve the inequality to get $x > 3$. 3 is the only number that is not greater than 3. Or, plug in 3 to get $11 < 11$. That's not true, so choose **A**.
4. **G.** $\dfrac{5}{11}$ Solve the inequality.
$$3x - 4 \leq -2x + 7$$
$$5x \leq 11$$
$$x \leq \dfrac{11}{5}$$

The smallest number is $\dfrac{11}{5}$, and its reciprocal is $\dfrac{5}{11}$.
5. **E.** $-p > -q$ Make up numbers that fit $0 < p < q$. For example, $p = 1$ and $q = 2$. Choice **E**, $-1 > -2$, is the only true answer.

Section 5.2 Absolute Value Equations and Inequalities
Practice Problems:
1. $x = 9$ or $x = 3$
2. $x < -4$ or $x > \dfrac{22}{3}$
3. $\dfrac{-1}{2} < x < \dfrac{7}{4}$
4. $x \leq -8$ or $x \geq 5$
5. $3 \leq x \leq 5$

5.2 ACT Problems:
1. **D.** $|x - 2| = 12$ Plug in 14 and –10.
2. **K.** –4   Solve to get $-4 < x < \dfrac{11}{2}$. –4 is not greater than –4. Or plug in answers.
3. **D.** $x \leq -3$ or $x \geq \dfrac{11}{5}$
$$5x + 2 = 13 \qquad 5x + 2 = -13$$
$$5x = 11 \qquad\quad 5x = -15$$
$$x = \dfrac{11}{5} \qquad\quad x = -3$$ The answer is

A or **D**. Plug in a number between them. 0 doesn't work, so it's an "or" problem. Choose **D**.
4. **H.** $y < 3$ Plug in $x = -3$.
$$|-3| > y$$
$$3 > y$$ Be careful not to choose G; the 3 is greater.
5. **D.** Only when $m = 0$ or $n = 0$ Try different numbers for m and n. For example, if $m = 0$ and $n = 2$, you get $|-2| = |2|$. That's true. If you switch m and n, it's also true.

Section 5.3 Solving Systems of Equations
Practice Problems:
1. $x = 1, y = 1$
2. $x = 2, y = -2$
3. $x = \dfrac{8}{3}, y = \dfrac{34}{3}$
4. $x = .5, y = 0$
5. $x = 2, y = 3$ or $x = .375, y = 16$
6. $d = 7, n = 8$

5.3 ACT Problems:
1. **E.** $(1, 1.5)$
$$2x + 4y = 8$$
$$-1(3x + 4y = 9)$$

$$2x + 4y = 8$$
$$\underline{-3x - 4y = -9}$$
$$-x = -1$$
$$x = 1$$

$$2(1) + 4y = 8$$
$$4y = 6$$
$$y = 1.5$$

116

2. **H.** $(0, -3)$

$$6x - 3y = 9$$
$$\underline{2x + 3y = -9}$$
$$8x = 0$$
$$x = 0$$

$$6(0) - 3y = 9$$
$$-3y = 9$$
$$y = -3$$

3. **C.** 0

$$3(2x + 3y = 1)$$
$$-2(3x + 5y = 2)$$

$$6x + 9y = 3$$
$$\underline{-6x - 10y = -4}$$
$$-y = -1$$
$$y = 1$$

$$2x + 3(1) = 1$$
$$2x = -2$$
$$x = -1$$ They asked for the sum of

the solutions. $1 - 1 = 0$.

4. **K.** 36

$$x + y = 15$$
$$\underline{x - y = 9}$$
$$2x = 24$$
$$x = 12$$

$$12 + y = 15$$
$$y = 3$$ They asked for xy. $12 \cdot 3 = 36$.

5. **D.** 3 Two equations have an infinite number of solutions when they are the same equation. Imagine graphing the same line twice. They touch at every point.

$$2(-2x + 3y = -9)$$
$$4x - 6y = 6a$$

$$-4x + 6y = -18$$
$$\underline{4x - 6y = 6a}$$
$$0 = -18 + 6a$$
$$18 = 6a$$
$$a = 3$$

Section 5.4 Matrices
Practice Problems

1. $\begin{bmatrix} 5 & 5 \\ 12 & 0 \end{bmatrix}$

2. $\begin{bmatrix} 26 & 60 \\ -49 & 7 \end{bmatrix}$

3. $\begin{bmatrix} -4 & 64 \\ 28 & 32 \end{bmatrix}$

4. $\begin{bmatrix} -96 & 0 \\ 24 & 40 \end{bmatrix}$

5. $\begin{bmatrix} 7 & 66 \\ 72 & 10 \end{bmatrix}$

6. $\begin{bmatrix} -4 & 4 \\ 4 & -2 \end{bmatrix}$

7. $\begin{bmatrix} 17 & -3 \\ 11 & -2 \end{bmatrix}$

8. 38

5.4 ACT Problems

1. **A.** $\begin{bmatrix} 0 & 21 \\ 18 & 11 \end{bmatrix}$ $2\begin{bmatrix} 3 & 5 \\ 4 & 6 \end{bmatrix} + \begin{bmatrix} -6 & 11 \\ 10 & -1 \end{bmatrix} = ?$

Multiply first. $\begin{bmatrix} 6 & 10 \\ 8 & 12 \end{bmatrix} + \begin{bmatrix} -6 & 11 \\ 10 & -1 \end{bmatrix} = ?$

Now add. $\begin{bmatrix} 6 & 10 \\ 8 & 12 \end{bmatrix} + \begin{bmatrix} -6 & 11 \\ 10 & -1 \end{bmatrix} = \begin{bmatrix} 0 & 21 \\ 18 & 11 \end{bmatrix}$

2. **G.** $\begin{bmatrix} 38 & -6 \\ 5 & 5 \end{bmatrix}$ $3\begin{bmatrix} 10 & -5 \\ 0 & 1 \end{bmatrix} + \begin{bmatrix} 8 & 9 \\ 5 & 2 \end{bmatrix} = ?$

Multiply first. $\begin{bmatrix} 30 & -15 \\ 0 & 3 \end{bmatrix} + \begin{bmatrix} 8 & 9 \\ 5 & 2 \end{bmatrix} = ?$

Now add. $\begin{bmatrix} 30 & -15 \\ 0 & 3 \end{bmatrix} + \begin{bmatrix} 8 & 9 \\ 5 & 2 \end{bmatrix} = \begin{bmatrix} 38 & -6 \\ 5 & 5 \end{bmatrix}$

3. **C.** $\$916$ On the t-shirts, they made $25 \times \$22 = \$550$, on the skirts, they made $18 \times \$12 = \$216$, and on the bracelets, they made $50 \times \$3 = \$150$. All together they made $\$550 + \$216 + \$150 = \$916$ Or use matrix multiplication.

$$[25 \; 18 \; 50] \times \begin{bmatrix} 22 \\ 12 \\ 3 \end{bmatrix} = 916$$

Section 5.5 Exponents and Logs
Practice Problems:

1. $9a^8 b^{13} c^{14}$

2. $2a^3 b^5$

3. $2^x \times (2^2)^x = 2^x \times 2^{2x} = 2^{3x}$

4. $\log_2 \left(\dfrac{3x^2}{x^2} \right) = \log_2(3)$

5. $\log(a^2) - \log(a^2) = 0$

117

6. 5

7. $x^2 = 36$

$x = 6$

8. $\log_5\left(\frac{81}{3}\right) = \log_5 x$

$x = 27$

9. $3^{(3x+1)} = (3^2)^{(x-1)}$

$3x + 1 = 2(x - 1)$

$x = -3$

10. $\log_2 10 - \log_2 5 = x$

$\log_2 2 = x$

$1 = x$

5.5 ACT Problems:

1. **B.** 1   Anything raised to the 0 equals 1.

2. **G.** 3 $b^{-3} = \frac{1}{27}$

$b^3 = 27$

$b = 3$

3. **C.** $\log\left(\frac{2xy^2}{3}\right)$

$\log x + \log y^2 + \log 2 - \log 3 = \log\left(\frac{2xy^2}{3}\right)$

4. **K.** 12 $\log_2(3 \times 4) = \log_2 x$

$\log_2 12 = \log_2 x$

$x = 12$

5. **D.** $3 < x < 4$ $2^x = 9$ We know that $2^3 = 8$ and $2^4 = 16$. 9 is between 8 and 19, so x is between 3 and 4. choose **D.**

Or use a calculator:

$\log_2 9 = \log(9)/\log(2) \approx 3.17$ Choose **D.**

Section 5.6 Rational and Radical Expressions
Practice Problems:

1. $x^{\frac{2}{3}} - x^{\frac{5}{3}} \div x^{\frac{1}{3}} + 1 = x^{\frac{2}{3}} - x^{\frac{5}{3} - \frac{1}{3}} + 1$

$x^{\frac{2}{3}} - x^{\frac{4}{3}} + 1$

2. $x^2 + \sqrt{x^{-1}} \cdot x^{\frac{1}{2}} = x^2 + x^{\frac{-1}{2}} \cdot x^{\frac{1}{2}}$

$x^2 + 1$

3. $\sqrt[2]{x^2} + \sqrt[3]{x^2} \cdot \sqrt[4]{x^2} \div \sqrt[5]{x^2} = x + x^{\frac{2}{3} + \frac{2}{4} - \frac{2}{5}}$

$x + x^{\frac{23}{30}}$

4. All real numbers

5. $x \neq 2$

6. $x \neq -2$ and $x \neq -3$

7. All real numbers

8. $\frac{3(x+y)(x-y)}{-(x+y)} = -3(x-y)$

$x + y \neq 0 \rightarrow x \neq -y$

9. $\frac{x(3-x)}{(x-3)(x+6)} = \frac{-x(x-3)}{(x-3)(x+6)} = \frac{-x}{x+6}$

$(x-3)(x+6) \neq 0 \rightarrow x \neq 3$ and $x \neq -6$

10. $\frac{(3x+2)(x+4)}{(x+2)(x+4)} = \frac{3x+2}{x+2}$

$x \neq -2, x \neq -4$

5.6 ACT Problems:

1. **E.** All nonnegative real numbers except 1

$x^2 - 1 \neq 0$

$(x+1)(x-1) \neq 0$

$x \neq 1$ or -1

Also, you can't have a negative in the square root, so all negatives have to be eliminated.

2. **F.** b must be positive   You can't have a square root in the radical, and a^2 will always be positive, so b has to be positive.

3. **C.** $1 - x$ $\frac{\sqrt[3]{x}}{\sqrt[3]{x}} - x^{\left(\frac{2}{3}\right)} \times x^{\left(\frac{1}{3}\right)} = 1 - x^1$

4. **H.** $\frac{x-5}{x+1}$ $\frac{(x-5)(x+1)}{(x+1)(x+1)} = \frac{(x-5)}{(x+1)}$ Pay no attention to $(x \neq 1)$. It's telling you that the expression is NOT undefined. Don't over complicate it.

5. **D.** $\frac{1}{x^2 - x}$ $\frac{x+1}{x(x+1)(x-1)} = \frac{1}{x(x-1)}$

They used a lot of words to say, "simplify this." Don't over complicate it.

Section 5.7 Complex Numbers
Practice Problems:

1. -12

2. $6i$

3. $-9 + 7i$

4. $4 - 2i$

5. $-2 - 14i$

6. $7 + i$

7. $31 + 29i$

8. $2 - i$

9. $-.6 + .2i$ or $\frac{-3 + i}{5}$

10. $-4 + 3i$

1. **B.** –2 $i^4 + 3i^2 = 1 + 3(–1) = –2$
2. **F.** –12 + 10i $–2(6 – 5i) = –12 + 10i$
3. **C.** –53 + 27i

$$(–5+2i)(11–i)=–55+5i+22i–2i^2$$
$$–55+27i–2(–1)=–53+27i$$

4. **J.** 24 + 10i $(5+i)(5+i)=25+5i+5i+i^2$
$$25+10i–1=24+10i$$

5. **B.** $\dfrac{3+37i}{13}$

$$\frac{9+5i}{2-3i}\times\frac{2+3i}{2+3i}=\frac{18+37i+15i^2}{4+6i-6i-9i^2}=\frac{3+37i}{13}$$

Section 5.8 Solving Quadratic Equations and Inequalities
Practice Problems:

1. $x=\pm 7$
2. $x<-5 \text{ or } x>2$
3. $x=1\pm 2i$
4. $-2\le x\le 2$
5. $\dfrac{-2-\sqrt{38}}{2}\le x\le\dfrac{-2+\sqrt{38}}{2}$
6. $x=-\dfrac{1}{2}\text{ or }x=-5$
7. $x=\pm i\sqrt{2}$
8. $x<0\text{ or }x>3$
9. $x=\dfrac{-9\pm\sqrt{73}}{4}$
10. $x\le-\dfrac{1}{100}\text{ or }x\ge\dfrac{1}{100}$

5.8 ACT Problems:

1. **A.** $-3<x<6$ $(x-6)(x+3)=0$, so $x = 6$ or –3. The answer is **A** or **B**. Choose a number between 6 and –3, like 0. Plug in: $(0)^2 – 3(0) – 18 < 0$ gives us –18 < 0. That's true, so choose the "x is between" answer, **A**.
2. **K.** –3i and 3i Solve for x: $x^2=-9$, so $x=\pm\sqrt{-9}=\pm 3i$ Or use the quadratic equation.
3. **B.** $x<-\sqrt{19}$ or $x>\sqrt{19}$ Solve for x: $x^2=19$, so $x=\pm\sqrt{19}$. The answer is **A** or **B**. Choose a number between $-\sqrt{19}$ and $\sqrt{19}$, like 0. Plug in: $(0)^2 > 19$. That's NOT true, so choose the "x is NOT between" answer, **B**.

4. **J.** 2   They are asking a simple question in a confusing way; don't assume it's hard. For the equation to have only one solution, we have to get the same answer twice. That means the two parentheses have to be the same. Now, just try to factor it. To start, only use the first and last terms: $x^2+k^2x+2k=(x+2)(x+k)$. For the parentheses to be the same, $k = 2$. Try plugging it in. $x^2+(2)^2x+2(2)=$ $x^2+4x+4=(x+2)(x+2)$ It works.
5. **B.** $-2x(x-4)(x+5)$ A cubic polynomial can factor just like a quadratic. To have solutions of 0, 4, and –5, you need parentheses of $(x-0)(x-4)$ and $(x+5)$. Eliminate answers **A** and **D**. To see if $f(1) =$ 36, plug in 1 for x. $–2(1)(1 – 4)(1 + 5) = 36$. Choose **B**.

Section 5.9 Sequences and Series
Practice Problems:

1. 4, 6, 8, 10, … 12, 14, 16
2. 100, 85, 70, 55, … 40, 25, 10
3. 8, 4, 2, … 1, $\dfrac{1}{2}, \dfrac{1}{4}$
4. $\dfrac{1}{3}$, –1, 3, –9, … 27, –81, 243
5. 1, 1, 2, 3, 5, 8, … 13, 21, 34 Add the two previous terms.

5.9 ACT Problems:

1. **A.** 4, 12, 36 $3\times 4=12, 12\times 3=36$
2. **H.** 21   It's a simple + 2 pattern and only 10 terms. Count on your fingers.
3. **E.** 38,880   The pattern is × 6: 3 × 6 = 30, 30 × 6 = 180. Multiply by 6 three more times to get the sixth term. 180 × 6 × 6 × 6 = 38,880.
4. **G.** 61   The first square has 4 hexagons. Each of the other squares adds 3 more hexagons. Start with 4 and add 3 19 times. 4 + 3(19) = 61.
5. **A.** 119.   The pattern is + 11. 108 + 11 = 119.

Intermediate Algebra Review:

1. B. $x \geq \dfrac{1}{2}$ $6x - 3(3x+2) \leq 5 - 5(3-x)$

$$6x - 9x - 6 \leq 5 - 15 + 5x$$
$$-3x - 6 \leq -10 + 5x$$
$$4 \leq 8x$$
$$\frac{1}{2} \leq x$$

2. G. $a_n = 6n + 6$ The recursive formula gives us the next number in the series based on the number before it. We know that a_1 (the first term) is 12. To find a_2 (the second term), we use the formula: $a_2 = a_1 + 6 = 12 + 6 = 18$. $a_3 = a_2 + 6 = 24$. Check the answers to see which one gives you 12, 18, 24, etc. Try **G**: $6(1) + 6 = 12$. $6(2) + 6 = 18$. $6(3) + 6 = 24$, etc.

3. A. $\dfrac{-x}{x+6}$

$$\frac{x(2-x)}{(x+6)(x-2)} = \frac{-x(x-2)}{(x+6)(x-2)} = \frac{-x}{x+6}$$

Ignore $x > 2$. It's just saying that the expression is NOT undefined.

4. H. $(-2, 7)$ $\begin{array}{l} 3x + 4y = 22 \\ -2(x + 2y = 12) \end{array}$

$$\begin{array}{r} 3x + 4y = 22 \\ -2x - 4y = -24 \\ \hline x \qquad\quad = -2 \end{array}$$

$$(-2) + 2y = 12$$
$$y = 7$$

5. A. 0 $x^2 - 5x + 4 = 0$
$$(x-1)(x-4) = 0$$
$x = 1$ and $x = 4$. This is an inequality, so the solutions are either everything between 1 and 4, or everything outside of 1 and 4. 2 and 3 are between the solutions, but there can't be two right answers. They both have to be wrong. 0 is the only number outside of 1 and 4. Choose **A**.

6. J. 2 $\left(x^{2a-1}\right)^4 = x^{12}$

$$x^{8a-4} = x^{12}$$
$$8a - 4 = 12 \qquad a = 2$$

7. E. $\dfrac{6x\sqrt{6x-y}}{6x-y}$

$$\frac{6x}{\sqrt{6x-y}} \times \frac{\sqrt{6x-y}}{\sqrt{6x-y}} = \frac{6x\sqrt{6x-y}}{6x-y}$$

That's a lot of words for an easy problem.

8. K. $\begin{bmatrix} 19 & 31 \\ 26 & 28 \end{bmatrix}$ $4\begin{bmatrix} 5 & 7 \\ 6 & 8 \end{bmatrix} + \begin{bmatrix} -1 & 3 \\ 2 & -4 \end{bmatrix} = ?$

Multiply first. $\begin{bmatrix} 20 & 28 \\ 24 & 32 \end{bmatrix} + \begin{bmatrix} -1 & 3 \\ 2 & -4 \end{bmatrix} = ?$

Now add. $\begin{bmatrix} 20 & 28 \\ 24 & 32 \end{bmatrix} + \begin{bmatrix} -1 & 3 \\ 2 & -4 \end{bmatrix} = \begin{bmatrix} 19 & 31 \\ 26 & 28 \end{bmatrix}$

9. C. $21 + 20i$ $(5+2i)^2 = (5+2i)(5+2i)$
$$25 + 5i + 5i + 4i^2 = 25 + 10i - 4$$
Don't forget that $i^2 = -1$

10. G. I and III only   Make up numbers. $|2+3| = 5$, $|2| + |3| = 5$, and $|-2-3| = 5$. All might work, so eliminate **K**. Before choosing **H**, try weird numbers. $|2+(-3)| = 1$, $|2| + |-3| = 5$, and $|-2-(-3)| = 1$. It's not **H**, it's **G**.

Chapter 6
Coordinate Geometry
9 questions

Section 6.1
Number Lines

Inequalities can be graphed on a number line. Shaded regions show what the variable can equal.

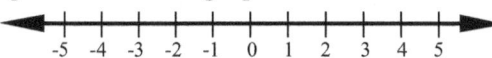

An open circle shows that the point is not included, such as $x > 3$.

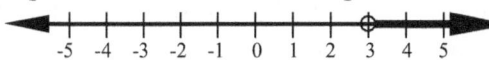

A closed circle shows that the point is included, for example, $x \leq 2$.

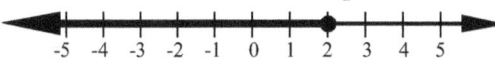

Example 1
Graph $x < 0$.

We put an open circle at 0 because x can't equal 0. The x is smaller than 0, so we shade the left side of the line.

Example 2
Graph $-4 \leq x < 3$.

We need a closed circle on -4 and an open circle on 3. The x is between -4 and 3, so we shade the area between the circles.

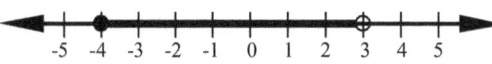

Example 3
Graph $2x^2 - 5x - 7 \geq 0$.

First, solve the quadratic. $\qquad\qquad (2x - 7)(x + 1) = 0$
$\qquad\qquad\qquad\qquad\qquad\qquad x = 3.5 \text{ and } x = -1$
Test a number between them. $\qquad 2(0)^2 - 5(0) - 7 \geq 0$
It does not work. $\qquad\qquad\qquad\qquad\qquad -7 \geq 0$
The answer is the outsides. $\qquad x \leq -1 \text{ or } x \geq 3.5$
Graph. Shade the outsides. Both numbers need closed circles.

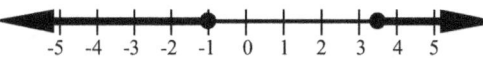

121

Graph each inequality on a number line.

1. $x > 2$

2. $x \leq 0$

3. $x \geq -1.5$

4. $2x + 3 > x + 6$

5. $x^2 \leq 4$

6.1 ACT Problems: Answers on p. 143

1. Which inequality is graphed below?

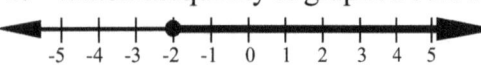

    A. $x > 2$
    B. $x < -2$
    C. $x \leq -2$
    D. $x \geq -2$
    E. $x \geq 2$

2. The graph below represents the solution of an inequality. Which of the following specifies the same solution in an algebraic form?

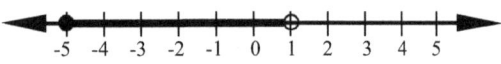

    F. $-5 = x = 1$
    G. $-5 < x < 1$
    H. $-5 < x \leq 1$
    J. $-5 \leq x < 1$
    K. $-5 \leq x \leq 1$

3. Which of the following is the graph of the solution set for $x^2 < 4$?

    A.

    B.
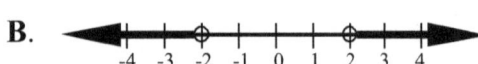

    C.

    D.

    E.

4. Which of the following shows the solution set for the inequality $5x + 1 > 11$?

    F.
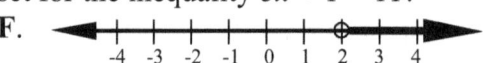

    G.
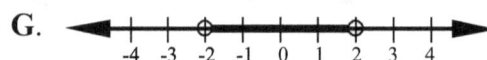

    H.
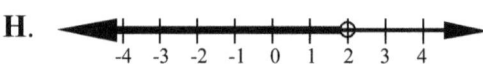

    J.
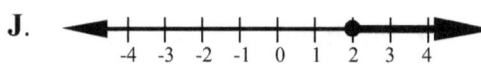

    K.
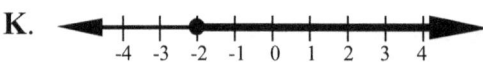

5. What is the sum of all integers that are solutions of the inequality graphed below?

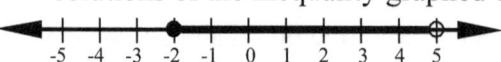

    A. 5
    B. 7
    C. 8
    D. 10
    E. 12

Section 6.2
Distance and Midpoint

Distance
The distance between two points can be found using the distance formula, $d=\sqrt{(x_2-x_1)^2+(y_2-y_1)^2}$, or using the Pythagorean Theorem, $a^2+b^2=c^2$.

Example 1
Find the distance between the points (–2, –5) and (4, 4).

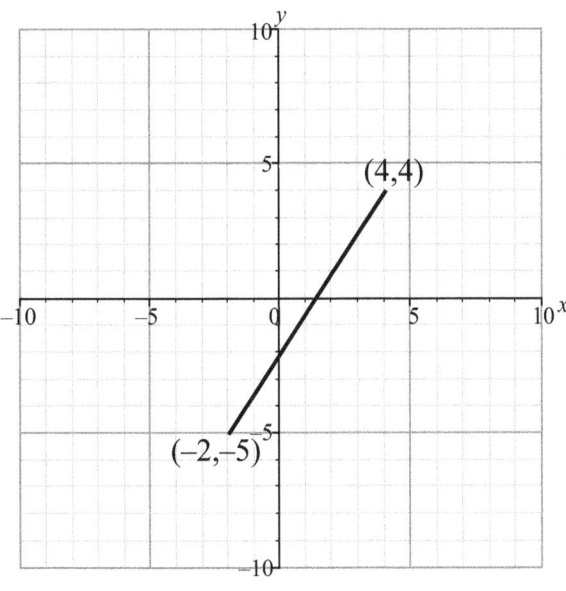

Method 1 – Distance Formula
If you remember the distance formula, just plug in the numbers.

$$d=\sqrt{(x_2-x_1)^2+(y_2-y_1)^2}$$
$$d=\sqrt{(4-(-2))^2+(4-(-5))^2}$$
$$d=\sqrt{(6)^2+(9)^2}$$

Simplify.

$$d=\sqrt{36+81}$$
$$d=\sqrt{117}=3\sqrt{13}$$

Method 2 – Pythagorean Theorem
Find the x distance and the y distance to substitute for a and b in the Pythagorean Theorem. The distance from –2 to 4 is 6, and the distance from –5 to 4 is 9. Look at the graph or imagine a number line if that helps. Plug in.

$$6^2+9^2=c^2$$

Solve.

$$117=c^2$$
$$c=\sqrt{117}=3\sqrt{13}$$

Method 3 – I can't remember anything
Most (not all, but most) drawings on the ACT are drawn to scale. If you could measure them, you could estimate the answer. Use a piece of scratch paper or your answer sheet to measure the length of the line segment. Next put the measurement against the axis to estimate the length. This line segment is between 10 and 11 units long. The correct answer is $\sqrt{117}\approx10.8$. Close enough.

Midpoint

The midpoint formula is $\left(\dfrac{x_1+x_2}{2}, \dfrac{y_1+y_2}{2}\right)$. If you don't remember the formula, average the x's and average the y's to get the same thing.

Example 2

Find the midpoint of the line segment joining the points (4,4) and (–2,–5).

These are the same points from the graph on the previous page.

To find the x coordinate of the midpoint, average the x values.

Do the same for the y values.

$$\frac{4+(-2)}{2}=\frac{2}{2}=1$$

$$\frac{4+(-5)}{2}=\frac{-1}{2}$$

The midpoint is $\left(1, -\dfrac{1}{2}\right)$. Plot this point on the graph to see that is is in the middle of the line segment.

Example 3

Find the length of the line segment with endpoints (–1, 0) and (2, 4).

Length and distance are the same thing. Find the distance between the points.

Distance Formula

$$d=\sqrt{(2-(-1))^2+(4-0)^2}$$
$$d=\sqrt{(3)^2+(4)^2}$$
$$d=\sqrt{9+16}=\sqrt{25}$$
$$d=5$$

Pythagorean Theorem

x-distance =3, y-distance=4
$$(3)^2+(4)^2=c^2$$
$$9+16=25=c^2$$
$$5=c$$

Example 4

Find the midpoint of the line segment with endpoints (–1, 0) and (2, 4).

Average the x values.

Average the y values.

Write the midpoint.

$$\frac{(-1)+2}{2}=\frac{1}{2}$$

$$\frac{0+4}{2}=\frac{4}{2}=2$$

$$\left(\frac{1}{2}, 2\right)$$

6.2 Practice Problems: Answers on p. 143

Find the length and the midpoint of the line segment with the following endpoints.
1. (0,0) and (6,8)
2. (2,4) and (10,4)
3. (–2,–3) and (2,3)
4. (1,5) and (2,9)
5. (–4,–3) and (5,9)
6. (–1,–3) and (2,3)
7. (10,12) and (1,–2)
8. (–8,–7) and (7,9)

1. In the standard (*x,y*) coordinate plane shown below, what is the distance in units from point *A* to point *B*?

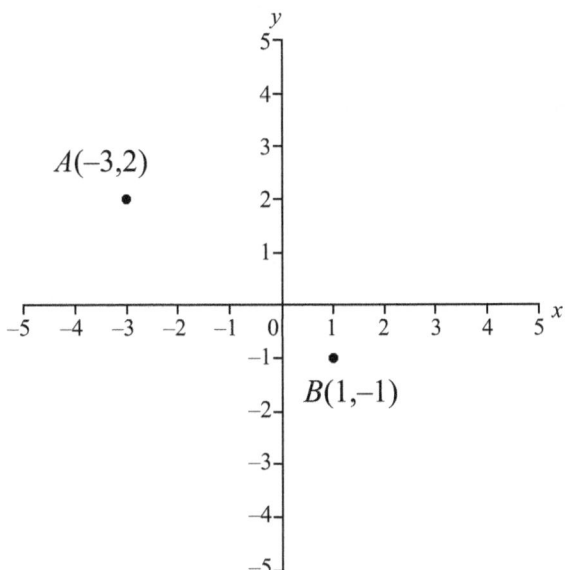

A. –5
B. –3
C. 3
D. 4
E. 5

2. One endpoint of a line segment in the (*x, y*) coordinate plane has coordinates (5, –8). The midpoint of the segment has coordinates (–1,4). What are the coordinates of the other endpoint of the segment?
F. (–7,16)
G. (–4,–32)
H. (2,–2)
J. (3,–6)
K. (4,12)

3. In the standard (*x,y*) coordinate plane, a right triangle has vertices at (–3,4), (–3,–4), and (3,–4). What is the length, in coordinate units, of the hypotenuse of this triangle?
A. 3
B. 4
C. 6
D. 8
E. 10

4. A park has the shape and dimensions in blocks given below. A water fountain is located halfway between *B* and *D*. If you are at *A*, the water fountain is:

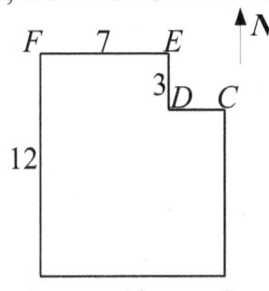

F. $4\frac{1}{2}$ blocks north and 5 blocks east.

G. $4\frac{1}{2}$ blocks north and $8\frac{1}{2}$ blocks east.

H. 6 blocks north and $3\frac{1}{2}$ blocks east.

J. 6 blocks north and 5 blocks east.

K. $7\frac{1}{2}$ blocks north and 9 blocks east.

5. Which of the following list the lengths of the sides of △*ABC* below from longest to shortest?

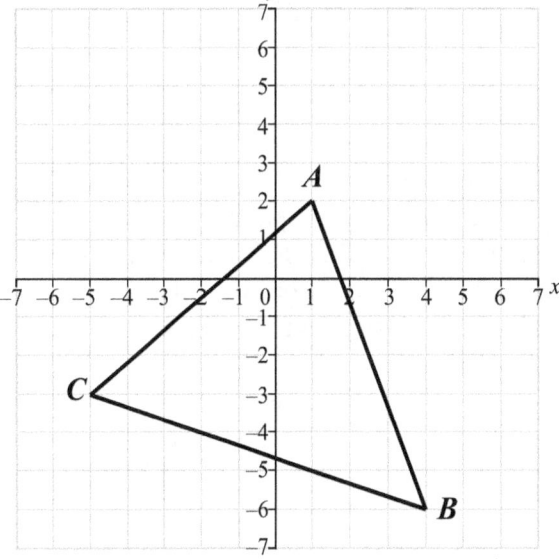

A. *AC, AB, BC*
B. *BC, AC, AB*
C. *BC, AB, AC*
D. *AB, AC, BC*
E. *AB, BC, AC*

125

Section 6.3
Graphing Linear Equations

Slope

$$m = \frac{y_2 - y_1}{x_2 - x_1} \quad \text{or} \quad slope = \frac{rise}{run}$$

If you follow a line from left to right, a positive slope goes up, but a negative slope goes down.

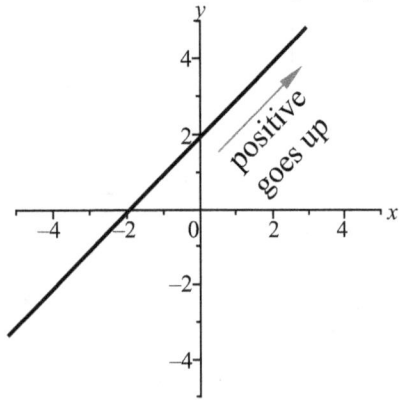 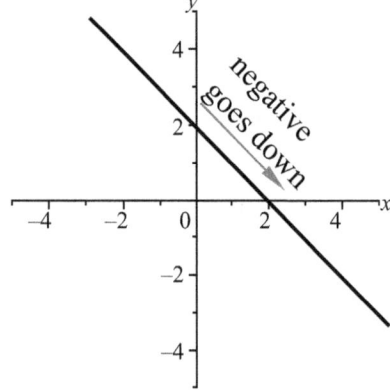

Example 1
Find the slope between the points (1,1) and (3,4).

If you remember the slope formula, plug in the numbers.

$$m = \frac{4-1}{3-1} = \frac{3}{2}$$

If you don't remember, choose one point to be the starting point. I'll use (1,1). To get from (1,1) to (3,4), how much do you rise? Rise is the y's, so to get from 1 to 4, you go up 3. The rise is +3. To get from (1,1) to (3,4), how much do you run? Run is x's, so to get from 1 to 3, you go right 2. The run is +2.

$$slope = \frac{rise}{run} = \frac{3}{2}$$

Plotting points
Any equation can be graphed by plotting points. Remember that the x coordinate is the horizontal axis, and it goes first. The y coordinate is vertical, and it goes second.

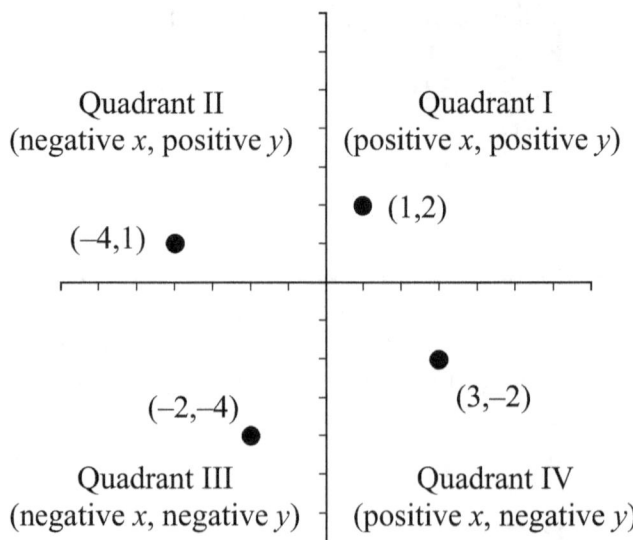

126

Example 2

Graph $2x + 3y = 12$.

First we plug in easy points. If $x = 0$, what is y?
$$2(0) + 3y = 12$$
$$3y = 12$$
$$y = 4$$
Plot the point (0,4)

If $y = 0$, what is x?
$$2x + 3(0) = 12$$
$$2x = 12$$
$$x = 6$$
Plot the point (6,0)

Connect the dots.

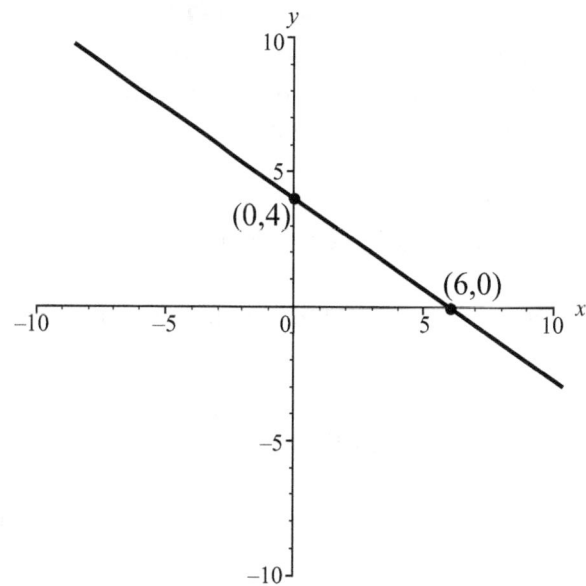

Slope – Intercept Form
Any linear equation (straight line) can be written in slope-intercept form.
$$y = mx + b$$
m = slope
b = y-intercept

Example 3

Graph $y = 2x - 3$

The equation is in slope-intercept form, so we don't need to rearrange it or plug in points.

The y-intercept is $b = -3$. This is where the line crosses the y-axis. Plot the point (0,–3).

The slope is $m = 2$, or $\dfrac{2}{1}$. That means the rise is +2 and the run is +1. Start at (0,–3) and go up 2 and right 1. That's the point (–1, 1). Plot the point.

You can go up 2 and over 1 as many times as you would like to get a more accurate line.

Connect the dots.

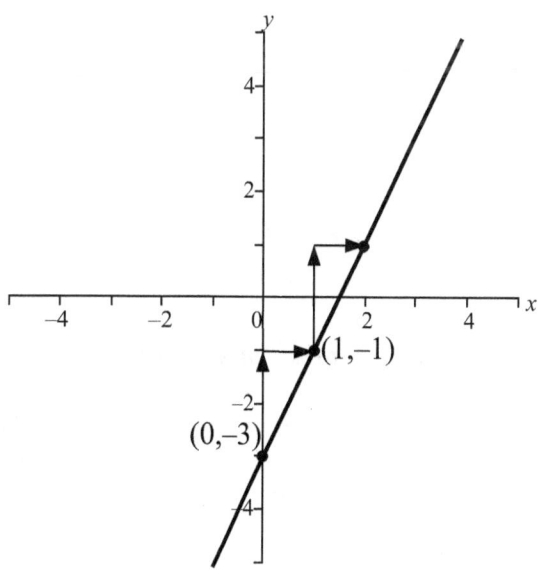

Parallel and Perpendicular Lines

Parallel lines have the same slope.

Perpendicular lines have slopes that are negative reciprocals ("upside-down and backwards" – flip the fraction and change the sign).

Example 4

What is the slope of a line that is perpendicular to the line with equation $2x+3y=6$?

First we want the equation in slope-intercept form.

$$3y=-2x+6$$
$$y=-\frac{2}{3}x+2$$

The slope of this line is $-\frac{2}{3}$.

$$m=-\frac{2}{3}$$

To get the perpendicular slope, we flip it over and change the sign.

$$m=\frac{3}{2}$$

6.3 Practice Problems: Answers on p. 144

Find the slope of the line passing through the following points.

1. $(1,1)$ and $(4,9)$
2. $(-2,0)$ and $(1,-5)$

Write the equation of the line with the following points or slopes.

3. $m=3, b=2$
4. $m=-1, (3,3)$
5. $m=\frac{2}{5}, (10,1)$
6. $(-2, 2)$ and $(0,3)$
7. $(-5, -1)$ and $(-1, 5)$

Graph the following equations

8. $y=4x-3$
9. $5x-3y=30$
10. $2x+7y=21$
11. $y=\frac{2}{3}x+5$
12. $y=1$

6.3 ACT Problems: Answers on p. 144-145

1. What is the slope of the line containing the points $(2,8)$ and $(-3, 10)$ in the standard (x,y) coordinate plane?

    A. $\frac{1}{18}$
    B. $\frac{12}{5}$
    C. 18
    D. $-\frac{5}{2}$
    E. $-\frac{2}{5}$

2. In the (x,y) coordinate plane, at which y-value does the line $x+2y=10$ intersect the y-axis?

    F. 1
    G. 2
    H. 5
    J. 10
    K. 12

128

3. What is the equation of a line with a y-intercept of 4 that is perpendicular to the line with equation $3x+5y=6$?

A. $y=-\dfrac{3}{5}x+4$

B. $y=-\dfrac{3}{5}x-4$

C. $y=\dfrac{5}{3}x+2$

D. $y=\dfrac{5}{3}x+4$

E. $y=\dfrac{5}{3}x-2$

4. The line $y=\dfrac{2}{3}x+2$ passes through which one of the following points?

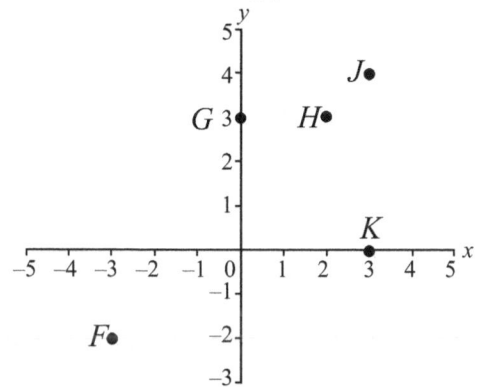

F. $F(-3,-2)$
G. $G(0,3)$
H. $H(2,3)$
J. $J(3,4)$
K. $K(3,0)$

5. Which equation is graphed below?

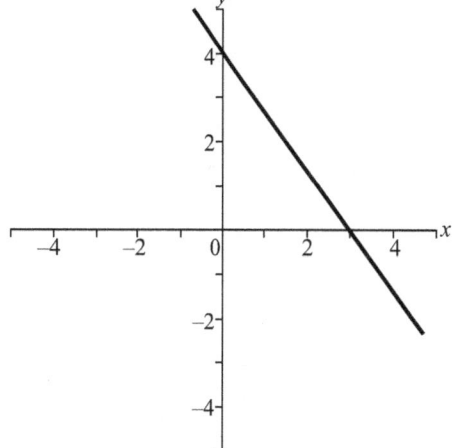

A. $y=\dfrac{3}{4}x+4$

B. $3x+4y=1$

C. $y=\dfrac{4}{3}x+3$

D. $4x-3y=4$

E. $y=-\dfrac{4}{3}x+4$

Section 6.4
Graphing Systems of Equations and Inequalities

Systems of Equations
Any system of linear equations can be solved by substitution, elimination, or graphing. Drawing a graph by hand is the *least* accurate method. If you are given two equations to solve, use substitution or elimination (see section 5.3). If you are given a graph, the solution is the intersection of the two lines.

There are three types of answers to these problems: one solution (a point), no solution, or infinitely many solutions.

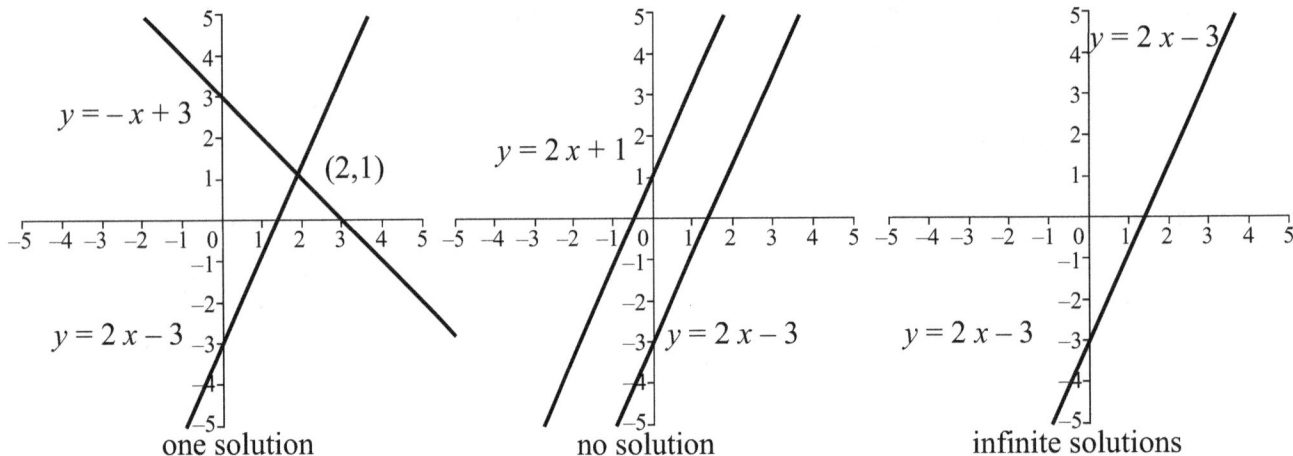

one solution   no solution   infinite solutions

A system of equations has one solution when the two lines are *not* parallel. This is the most common type. A system has no solution when the lines are parallel with different y-intercepts because the lines never touch. The system has infinitely many solutions when you have the same line twice because the lines touch each other at every point.

ACT loves the wording of "infinite solutions" because it sounds confusing. *Infinite solutions just means that you have the same line twice*.

Systems of Inequalities

To graph an inequality, first graph the line. On a number line, $<$ and $>$ have an open circle. On a graph, they have a dotted line. On a number line $\leq$ and $\geq$ have a closed circle, and on a graph they have a solid line. On multiple choice questions, check these easy things first.

Next, we shade one side of the line. The easiest way to check which side is shaded is to pick a point that is NOT on the line, and plug it in to the inequality. If it works, shade that side. If it doesn't, don't shade that side. Shade the other side instead.

To graph a system of inequalities, graph them one at a time and lightly shade the correct side of each line. The area that is shaded twice is the solution.

Example 1
Graph the inequality $y < x + 2$.

First, just graph the line. The y-intercept is 2 and the slope is 1.
Next, make it a dotted line because of the $<$ sign.
Finally, plug in a point that is *not* on the line, like (0,0)

$\qquad 0 < 0 + 2$
$\qquad 0 < 2$

That's true, so shade towards the point (0,0).

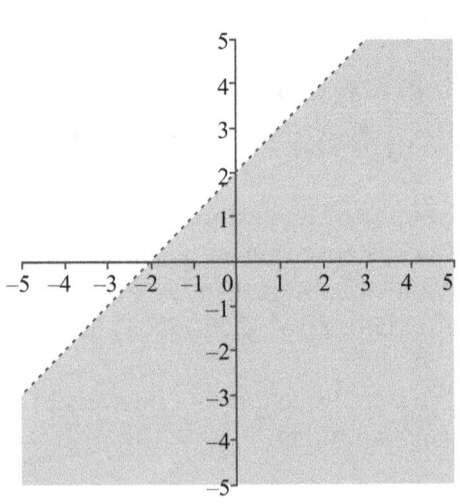

130

Example 2

Graph $3x + y > 6$ and $2x - 3y \le 6$

First graph the lines either by plugging in points or using slope-intercept form.

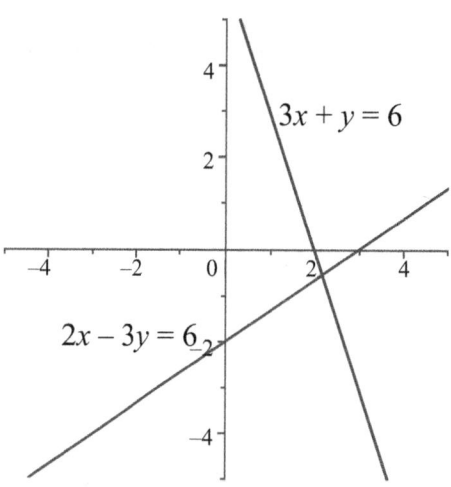

$3x + y = 6 \rightarrow y = -3x + 6$
The y-intercept is 6 and the slope is -3.

$2x - 3y \le 6$
$2(0) - 3y = 6 \rightarrow y = -2$
$2x - 3(0) = 6 \rightarrow x = 3$
Plot the points $(0,-2)$ and $(3,0)$. Connect the dots.

Make the first line a dotted line.
Plug in $(0,0)$.
    $3(0) + (0) > 6$
    $0 > 6$
False, shade the side away from $(0,0)$.

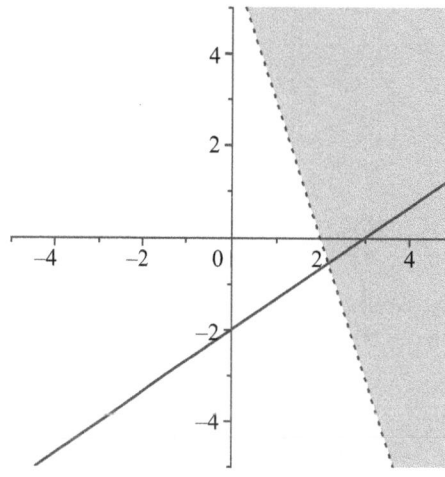

Keep the second line solid because the points on the line work in the inequality.
Plug in $(0,0)$
    $2(0) - 3(0) \le 6$
    $0 \le 6$
True, shade toward $(0,0)$.

The solution set is the region that is shaded twice.

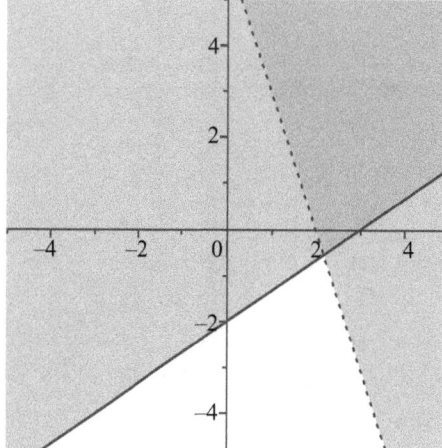

6.4 Practice Problems: Answers on p. 145

Graph each system of inequalities.

1. $y > \dfrac{1}{2}x - 4$ and $y \le 3x - 1$

2. $y \ge -x + 2$ and $y \le 2x - 2$

3. $2x - 5y > 5$ and $2y - 5x \ge 4$

4. $x + y > 1$ and $x + y < 6$

131

1. Which is NOT a solution of the inequality
   $4x+3y>10$?
   A. (3,2)
   B. (0,3)
   C. (2,3)
   D. (3,4)
   E. (3,0)

2. Which inequality is graphed below?

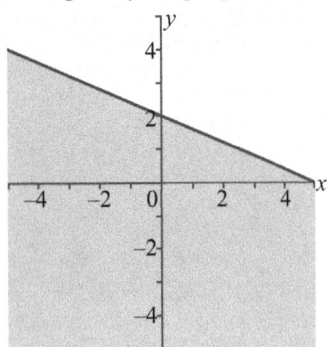

   F. $2x+5y\leq10$
   G. $2x-5y\leq10$
   H. $5x-2y\leq10$
   J. $5x+2y\geq10$
   K. $5y-2x\leq10$

3. Which system of inequalities is graphed below?

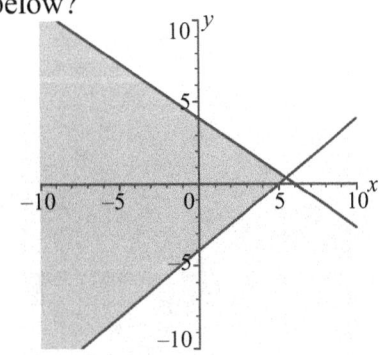

   A. $2x+3y\geq12$
      $4x-5y\leq20$

   B. $2x+3y\geq12$
      $4x+5y\leq20$

   C. $2x+3y\leq12$
      $4x-5y\leq20$

   D. $2x+3y\leq12$
      $4x+5y\geq20$

   E. $2x+3y\leq12$
      $4x-5y<20$

4. Which equation creates an infinite number of solutions when solved in a system with
   $y=6x+7$?
   F. $y=7x+6$
   G. $12x+2y=14$
   H. $12x+3y=21$
   J. $18x-3y=-21$
   K. $2y-12x=21$

5. Which is the graph of the inequality
   $2x-y>-3$?
   A.

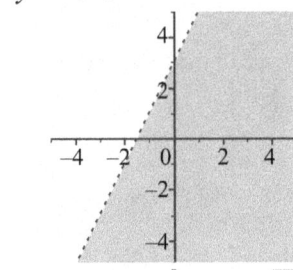

   B.

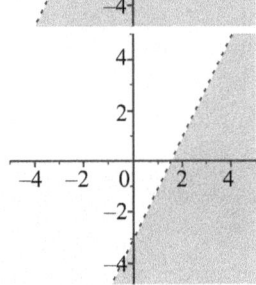

   C.

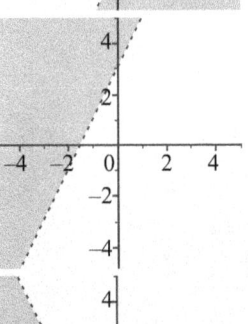

   D.

   E.

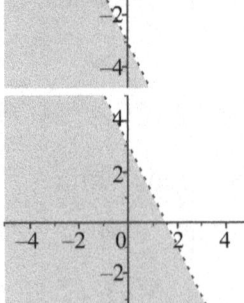

132

Section 6.5
Graphing Conic Sections

Conic sections are curves made from sections of a cone.

The most commonly tested conic section on the ACT is the circle. Parabolas are sometimes seen, but the ellipse and hyperbola are more rare.

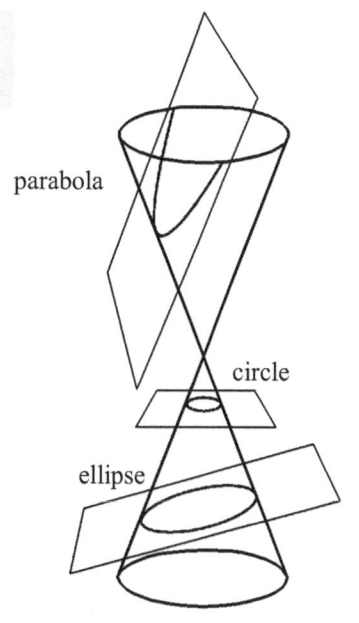

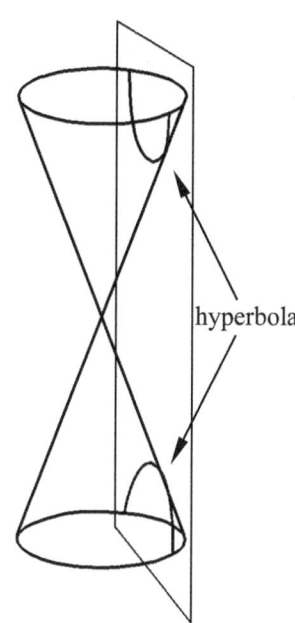

Circle

The equation of a circle is a cross between the Pythagorean Theorem and factoring. The standard form for the equation of a circle is

$$(x-h)^2+(y-k)^2=r^2$$

The center is (h,k) and the radius is r.

Example 1

The circle at right has equation $(x-1)^2+(y-2)^2=4$. Find the center and the radius.

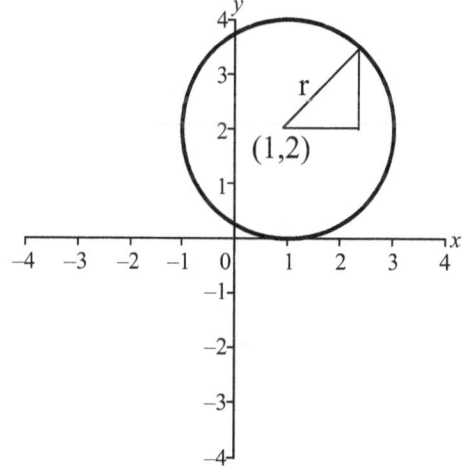

You can make a right triangle using any point on the circle and the center, like the picture to the right. The Pythagorean Theorem gives us $x^2+y^2=r^2$ using the x and y distances. Looking at our equation, we know that $r^2 = 4$, so $r = 2$.

Factoring helps us find the center. If we had factored a polynomial and had gotten $(x - 1)$ as one of the parentheses, we would set it equal to zero and solve for x.

$$(x-1)=0 \quad (y-2)=0$$
$$x=1 \qquad y=2$$

The center is (1,2) and the radius is 2.

Parabola

The equation of a parabola has either an x^2 or a y^2, but not both. The common forms of the equation are

$$y=ax^2+bx+c$$
$$x=ay^2+by+c$$
$$y=a(x-h)^2+k$$

The factored form can be used to find the coordinates of the vertex, (h,k). Otherwise, plug in enough points to identify the graph or use a graphing calculator.

Example 2

Graph the equation $y = 2x^2 + 4x$.

If you remember factoring by completing the square, you can rearrange the equation to find the vertex. $y = 2(x+1)^2 - 2$ or $y + 2 = 2(x+1)^2$ The vertex is $(-1,-2)$. Plug in points to finish the graph.

If you don't remember completing the square, just start plugging in points.

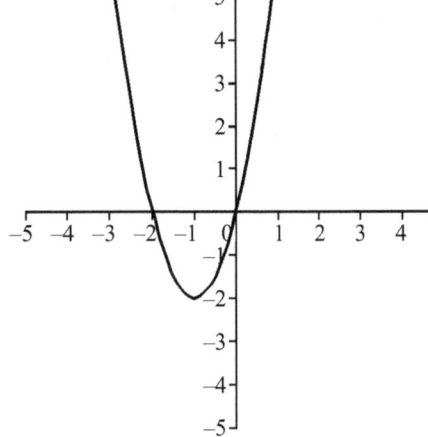

x	y
-2	0
-1	-2
0	0
1	2
2	16

Because -2 and 0 both have y-coordinates of 0, the vertex must be exactly between them. Therefore $(-1,-2)$ is the vertex. Plot the points (except for (2,16), the 16 is too big) and connect the dots.

Ellipse and Hyperbola

Lets be honest, no one really remembers ellipse and hyperbola. An ellipse is an oval, and a hyperbola looks like two back-to-back parabolas. Their equations are similar to circles with different radii for x and y.

Ellipse
$$\frac{(x-h)^2}{a^2} + \frac{(y-k)^2}{b^2} = 1$$

Hyperbola
$$\frac{(x-h)^2}{a^2} - \frac{(y-k)^2}{b^2} = 1$$

Both have centers at (h,k). Both have a "radius" of a in the x direction and a "radius" of b in the y direction. The ellipse has a plus sign between the terms, and the hyperbola has a minus. This hyperbola is horizontal because the x term comes first. If the y term comes first, the hyperbola will be vertical.

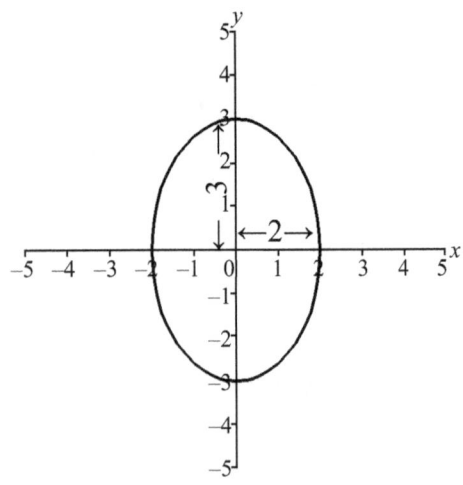

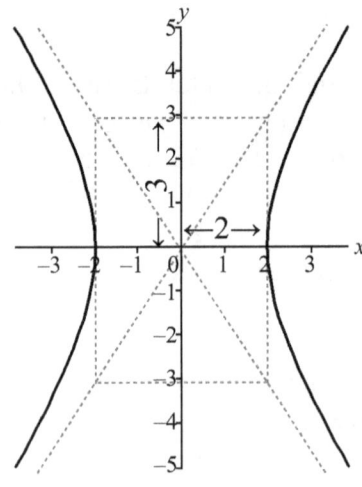

$$\frac{x^2}{2^2} + \frac{y^2}{3^2} = 1$$

$$\frac{x^2}{2^2} - \frac{y^2}{3^2} = 1$$

Both graphs have centers at $(h,k) = (0,0)$, x-radius $a = 2$, and y-radius $b = 3$.

134

6.5 Practice Problems: Answers on p. 146

Graph the following equations.

1. $(x+1)^2+(y-1)^2=25$
2. $(x-2)^2+y^2=17$
3. $y=x^2-5x+6$
4. $\dfrac{(x-1)^2}{4}+\dfrac{(y-1)^2}{16}=1$
5. $\dfrac{(y-1)^2}{4}-\dfrac{(x-1)^2}{16}=1$

6.5 ACT Problems: Answers on p. 146-147

1. A particular circle in the standard (x,y) coordinate plane has an equation of $(x-5)^2+y^2=13$. What are the radius of the circle, in coordinate units, and the coordinates of the center of the circle?

	radius	center
A.	6.5	(−5,0)
B.	6.5	(5,0)
C.	13	(5,0)
D.	$\sqrt{13}$	(−5,0)
E.	$\sqrt{13}$	(5,0)

2. The graph of the equation $h=-at^2+bt+c$, which describes how the height, h, of a thrown football changes over time, t, is shown below.

   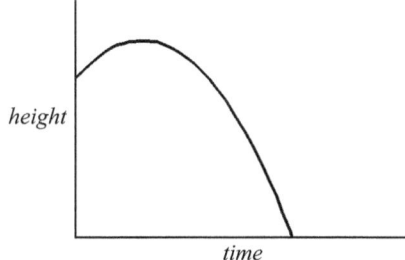

   If you alter only this equation's c term, which gives the height at time $t = 0$, the alteration has an effect on which of the following?

   I. The h-intercept
   II. The maximum value of h
   III. The t-intercept
   F. I only
   G. II only
   H. III only
   J. I and III only
   K. I, II, and III

3. Which is the equation of an ellipse?
   A. $x^2+2y^2=16$
   B. $x^2-y^2=9$
   C. $(x+1)^2+(y+1)^2=25$
   D. $4x^2+4y^2=8$
   E. $y=2x^2+5x+3$

4. One of the following is an equation for a circle centered at C and passing through the origin in the standard (x,y) coordinate plane below. Point C is 4 coordinate units from the y-axis and 3 from the x-axis. Which equation determines the circle?

   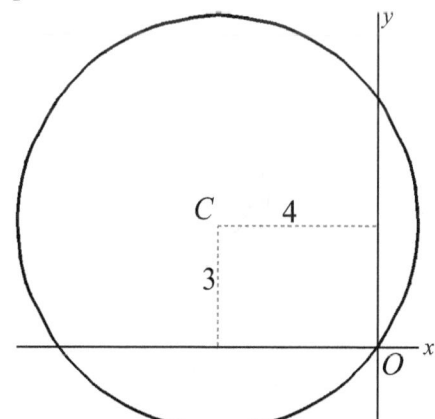

   F. $(x+3)^2+(y+4)^2=25$
   G. $(x+3)^2+(y-4)^2=25$
   H. $(x+4)^2+(y-3)^2=25$
   J. $(x-3)^2+(y+4)^2=25$
   K. $(x-4)^2+(y+3)^2=25$

135

5. Which is the graph of
$$9y^2 - 4x^2 - 18y - 16x = 43 \ ?$$

A.

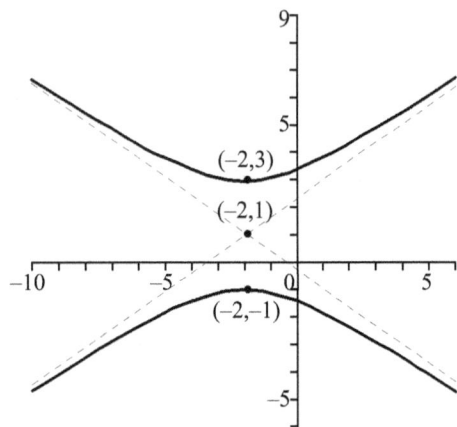

D.

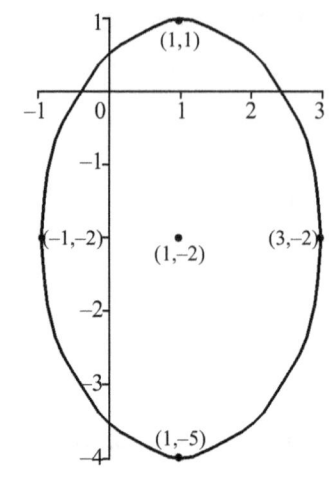

B.

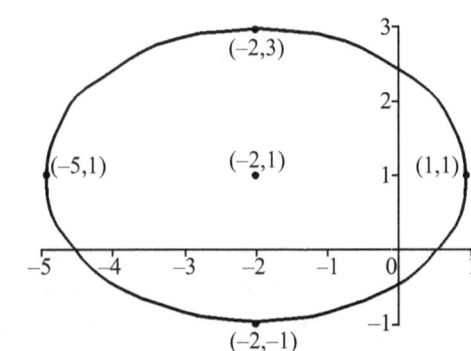

E.

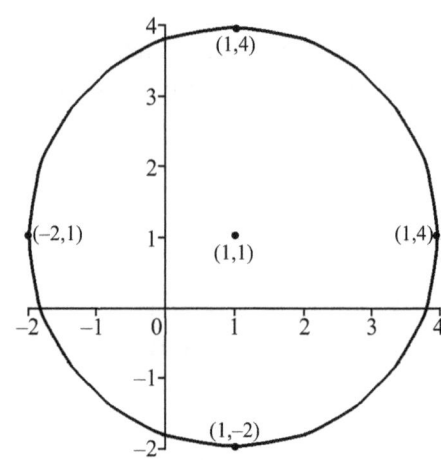

C.

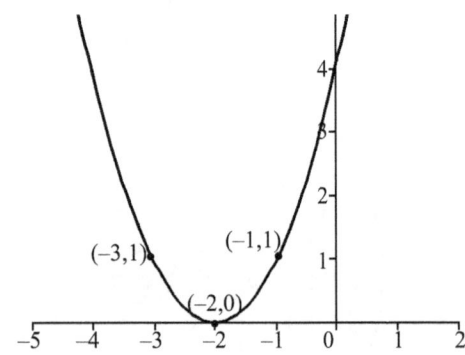

Section 6.6
Transformations

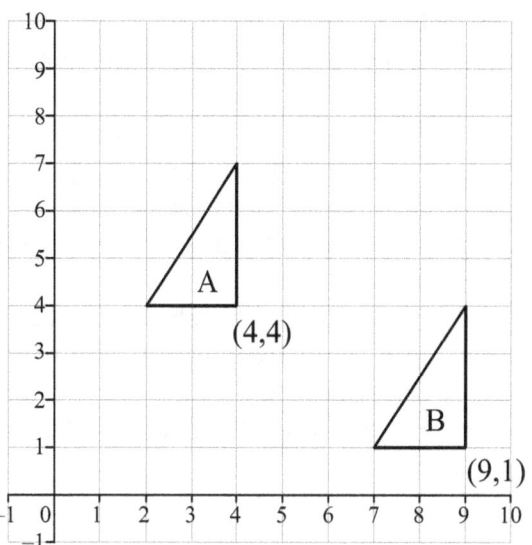

Translations

"Translate" is a technical term for "move." The ACT loves to use big words for simple ideas.
The graph to the right shows that triangle A was translated 5 units right and 3 units down to create triangle B.

If you don't have a picture, add or subtract from the x and y coordinates.

5 units right means add to the x.
$4 + 5 = 9$
3 units down means subtract from the y.
$4 - 3 = 1$

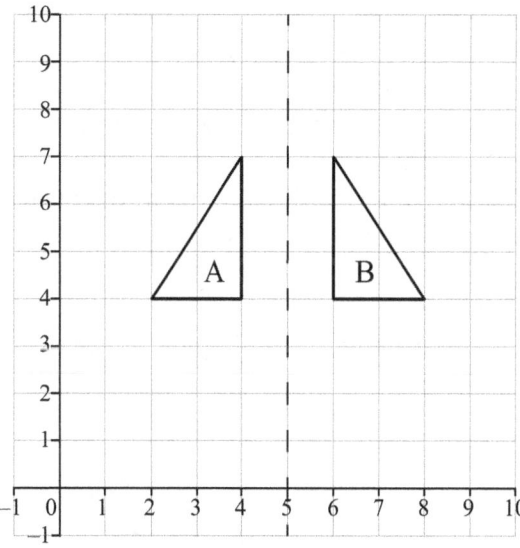

Reflections

When you look in the mirror, you see your reflection.
The graph to the right shows triangle A reflected across the line $x = 5$ to create triangle B.

If you stand 1 step from a mirror, your reflection is standing 1 step from the mirror. Triangle A is 1 unit from $x = 5$, and triangle B is 1 unit from $x = 5$.

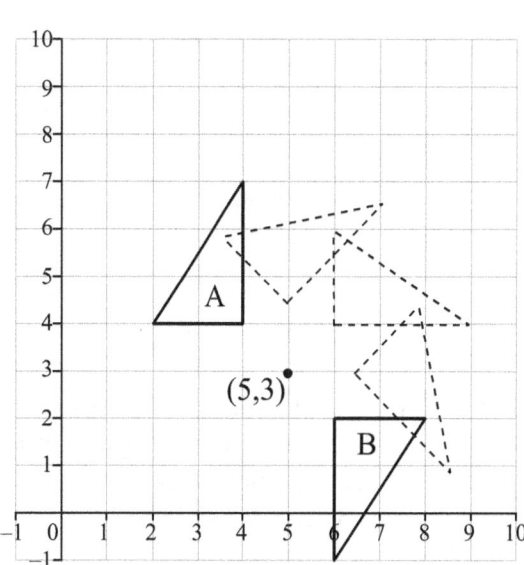

Rotations

Rotate means turn.

The graph to the right shows triangle A rotated 180° about the point (5,3).

Put a piece of scratch paper over the graph, trace the figure, and hold the tip of your pencil on the point you want to rotate around. Turn the paper and see where your drawing moves.

Example 1

Which of the following transformations can be used to produce rectangle B from rectangle A?

A. Horizontal translation right 11 units followed by a vertical translation up 3 units
B. Reflection across the y axis
C. Reflection across the y axis followed by a vertical translation up 3 units
D. Horizontal translation right 2 units, followed by a reflection across the y axis, followed by a vertical translation up 3 units
E. 180° rotation about the origin

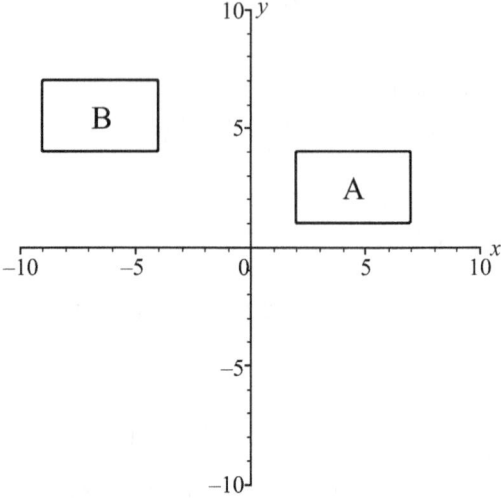

Start by checking the easiest ones. A 180° rotation about the origin would put the rectangle in the third quadrant. Eliminate **E**. Rectangle A needs to move to the left and up, not to the right. Eliminate **A**. Reflecting across the y axis would not move it up. Eliminate **B**.
To check the more complicated ones, follow one point. The top left corner starts at (2,4).
Answer **C** reflects it to (–2,4) and then shifts it up to (–2,7). Rectangle B does not have a corner at (–2,7). Eliminate **C**.
Answer **D** shifts it to (4,4) then reflects it to (–4,4), and then shifts it up to (–4,7). Rectangle B does have a corner at (–4,7). Choose **D**.

Example 2

Create rectangle B by flipping rectangle A across the x axis and then flipping it across the y axis.

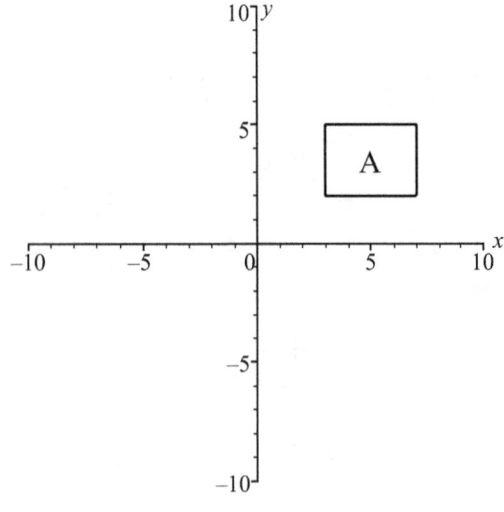

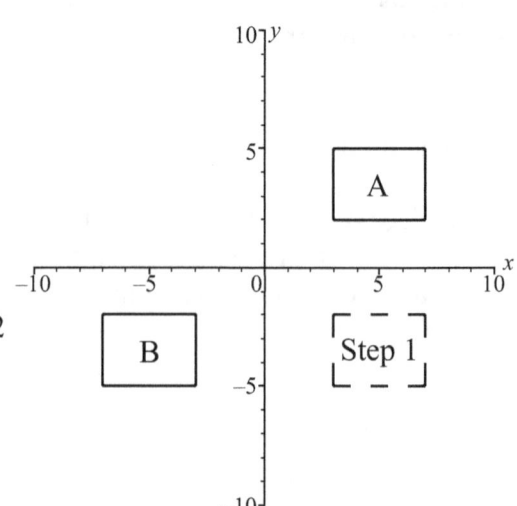

Step 1: flip the rectangle down over the x axis. Rectangle A is 2 units above the axis, your rectangle should be 2 units below. Next flip it left over the y axis. You start 3 units to the right, so rectangle B is 3 units to the left.

6.6 Practice Problems: Answers on p. 147

Describe the transformation used to produce triangle B from triangle A.

1.

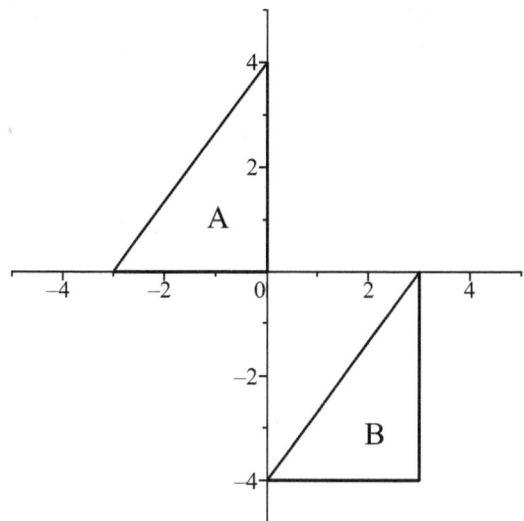

2.

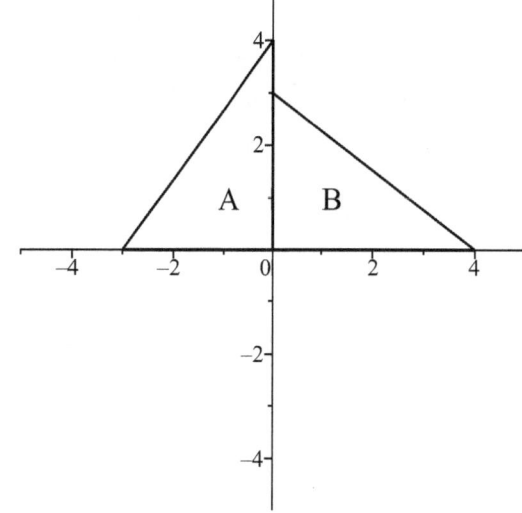

3.

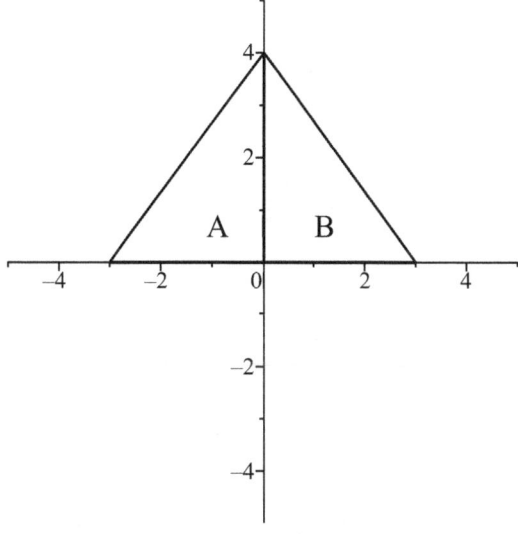

4. Create triangle B by shifting triangle A right 4 units.

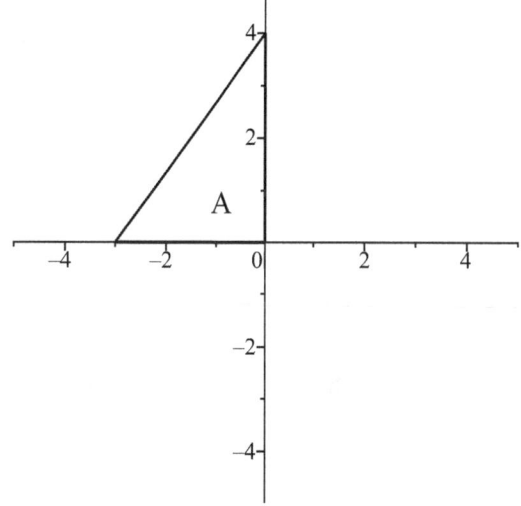

5. Create triangle B by rotating triangle A 180° about the point (−1.5, 2).

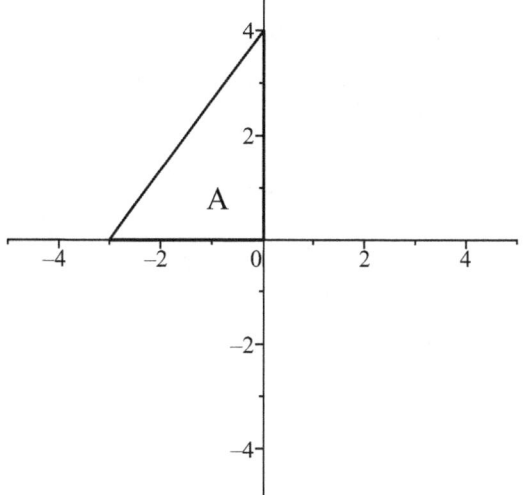

6. Create triangle B by flipping triangle A over the line $y = x$.

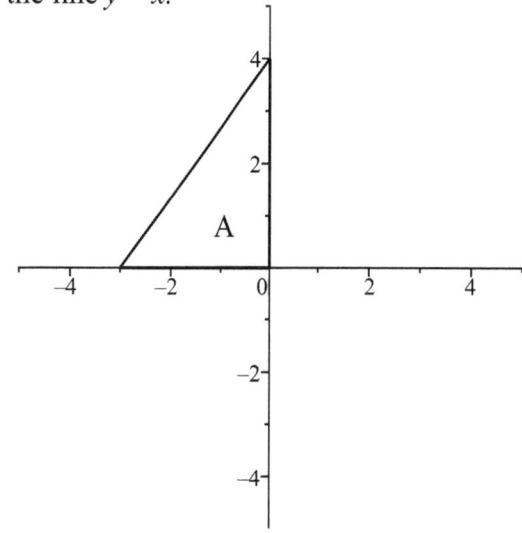

1. Figure *ABCDE*, shown in the standard (*x,y*) coordinate plane below, has been reflected across a line to figure *A'B'C'D'E'*. Which of the following lines of reflection was used for this transformation?

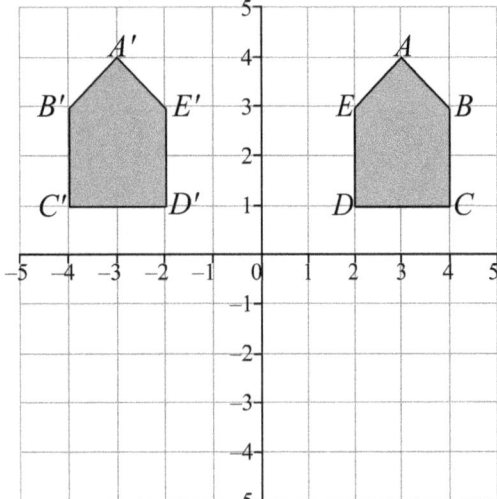

   A. *y* = 0
   B. *x* = 0
   C. *y* = 1
   D. *y* = *x*
   E. *y* = −*x*

2. Square A in the figure below is translated 2 units right and 4 units down to create square A'. Which of the following shows the coordinates of a vertex of square A'?

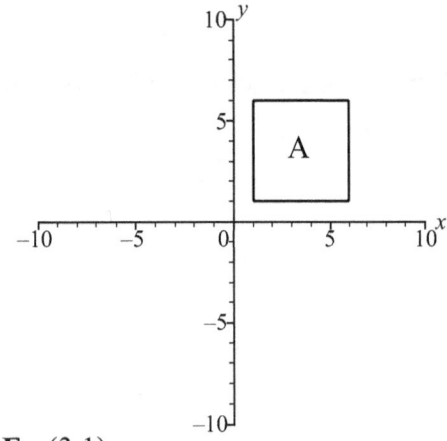

   F. (3,1)
   G. (3,2)
   H. (3,3)
   J. (3,4)
   K. (3,5)

3. The graph of $f(x) = (x - 1)^2$ is shown below. $f'(x)$ is the reflection of $f(x)$ across the *x* axis. Which of the following shows the graph of $f'(x)$?

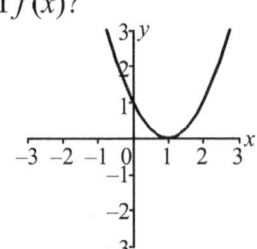

   A.

   B.

   C.

   D.

   E.

4. The triangle, $\triangle ABC$, shown below has side lengths as labeled. A' is the image of point *A* when the triangle is reflected across $\overline{BC}$. What is the perimeter of quadrilateral *A'BAC*?

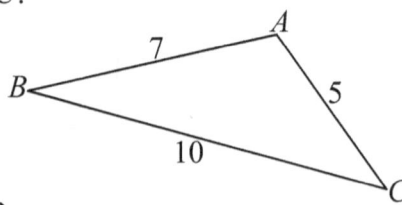

   F. 12
   G. 22
   H. 24
   J. 30
   K. 34

Coordinate Geometry Review

1. Which is the equation for the graph shown below?

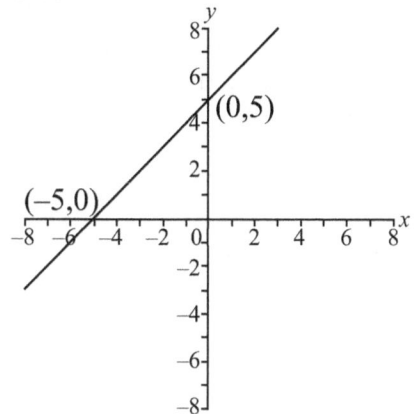

   A. $y=-5x+5$
   B. $y=x+5$
   C. $y=-x+5$
   D. $y=-5x+1$
   E. $y=5x+1$

2. What is the midpoint of a line segment whose endpoints are (1, 10) and (8, −2)?
   F. (4.5, 4)
   G. (4, 4.5)
   H. (3.5, 6)
   J. (5, 4)
   K. (4.5, 4.5)

3. What is the equation of a line that passes through the point (3,4) and is perpendicular to the line with equation $2y+3x=8$?
   A. $y=\dfrac{3}{2}x+2$

   B. $y=-\dfrac{3}{2}x+4$

   C. $y=\dfrac{2}{3}x+2$

   D. $y=-\dfrac{2}{3}x+6$

   E. $y=\dfrac{2}{3}x+4$

4. Which of the following shows the solution set for the inequality $6x+7\geq-5$?

   F.
   G.
   H.
   J.
   K.

5. In the coordinate plane, what is the distance between the points with coordinates (−1,−1) and (−6, 4)?
   A. $\sqrt{10}$
   B. 5
   C. $5\sqrt{2}$
   D. 10
   E. 25

6. What is the slope of the line through (−3,8) and (−5,3) in the standard (x,y) coordinate plane?
   F. $\dfrac{5}{2}$

   G. $-\dfrac{5}{2}$

   H. $\dfrac{2}{5}$

   J. $-\dfrac{2}{5}$

   K. $\dfrac{5}{3}$

7. $y=2x^2+7x+6$ is the equation of
   A. a circle
   B. a parabola
   C. a line
   D. an ellipse
   E. a hyperbola

141

8. For 6 hours a hose is used to fill a pool up to the 2-foot mark. A second hose with a faster flow rate is then added, and both hoses are used to fill the pool up to the 5-foot mark. Which of the following graphs shows the relationship between the time spent filling the pool and the height of the water in the pool?

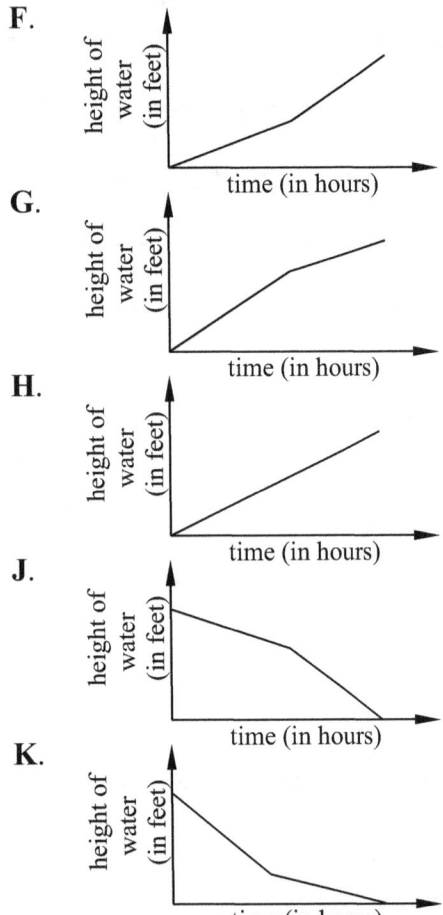

9. Quadrilateral *QUAD* has vertices (–3, 5), (2, 4), (3, 0), and (–4, –1) in the standard (*x, y*) coordinate plane. If *QUAD* is translated 2 units to the right and 3 units up to form quadrilateral *Q'U'A'D'*, which of the following shows the coordinates of the vertices of *Q'U'A'D'*?

A. (–5, 8), (4, 7), (5, 0), (–6, –4)
B. (–1, 8), (4, 7), (5, 3), (–2, 2)
C. (0, 7), (5, 6), (6, 2), (–1, 1)
D. (–2, 6), (3, 5), (4, 1), (–3, 0)
E. (5, –3), (4, 2), (0, 3), (–1, –4)

10. Tickets for the Student Talent Show Fund Raiser at Quest Elementary School are $5 for adults and $3 for students. To cover expenses, a total of $300 must be collected from ticket sales for the show. Which of the following graphs in the standard (*x, y*) coordinate plane, where *x* is the number of student tickets sold and *y* is the number of adult tickets sold, represents all the possible combinations of ticket sales that cover at least $300 in expenses?

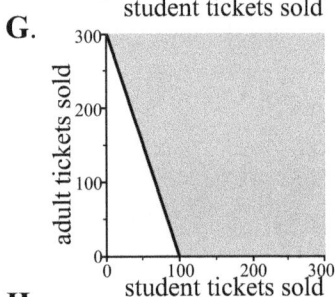

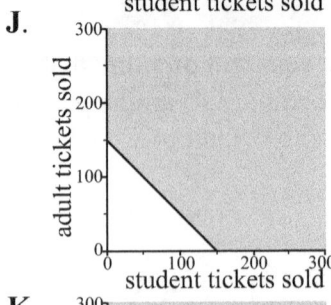

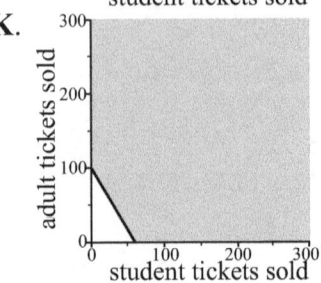

Answers on p. 148

Answers and Explanations

Section 6.1 Number Lines
Practice Problems:

1.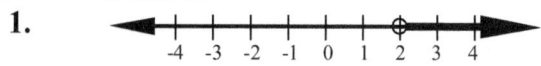

2.

3.

4.

5.

6.1 ACT Problems:

1. **D.** $x \geq -2$ The circle is on –2; eliminate **A** and **E**. The numbers bigger than –2 are shaded, so it's "greater than." Eliminate **B** and **C**. Also, it's a solid circle; **D** is definitely correct.

2. **J.** $-5 \leq x < 1$ The –5 has a closed circle, so it needs $\leq$. The 1 has an open circle, so it needs $<$. Choose **J**.

3. **A.**

    First, solve for x. $x^2 = 4 \rightarrow x = \pm 2$. The circles go on 2 and –2; eliminate **E**. The sign was $<$, so we need open circles. Eliminate **C** and **D**. Plug in a number between 2 and –2. $(0)^2 < 4$. True, shade the zero. Choose **A**.

4. **F.**

    First, solve for x.
    $5x + 1 > 11 \rightarrow 5x > 10 \rightarrow x > 2$. We need an open circle on 2, and x is greater, so we shade the bigger numbers.

5. **B.** 7 Add the shaded integers. Be careful, 5 is not shaded. $-2 + -1 + 0 + 1 + 2 + 3 + 4 = 7$

Section 6.2 Distance and Midpoint
Practice Problems:

1. d = 10,   mp = (3, 4)
2. d = 8,   mp = (6, 4)
3. d = $2\sqrt{13}$, mp = (0, 0)
4. d = $\sqrt{17}$, mp = (1.5, 7)
5. d = 15,   mp = (0.5, 3)
6. d = $3\sqrt{5}$, mp = (0.5, 0)
7. d = $\sqrt{277}$, mp = (5.5, 5)
8. d = $\sqrt{481}$, mp = (–0.5, 1)

6.2 ACT Problems:

1. **E.** 5 x-distance = 4, y-distance = 3.
    $4^2 + 3^2 = d^2 \rightarrow d = 5$ Don't forget that distance is always positive.

2. **F.** (–7,16) In the x-direction, to get from the endpoint, 5, to the midpoint, –1, subtract 6. To get from –1 to the other endpoint, subtract 6 again. $-1 - 6 = -7$. Choose **F**.

3. **E.** 10 Sketch a graph of the triangle. The x-distance is 6, and the y-distance is 8. Use the Pythagorean Theorem to find the hypotenuse. $6^2 + 8^2 = c^2$ $c = 10$.

    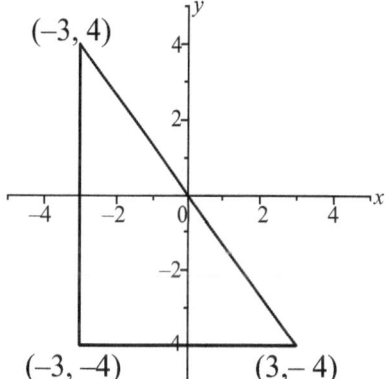

4. **G.** $4\frac{1}{2}$ blocks north and $8\frac{1}{2}$ blocks east.

    We need to get to a point halfway between B and D, starting from A. We'll make A the origin and then find the coordinates of B and D by subtracting the given distances. Find the midpoint of B and D.
    $\frac{10+7}{2} = 8\frac{1}{2}$ and
    $\frac{0+9}{2} = 4\frac{1}{2}$.

    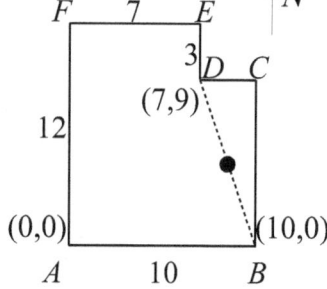

    Choose G. Notice that they put the north direction first. That's y, not x.

5. **C.** *BC, AB, AC* Use the edge of a piece of paper to measure.

Section 6.3 Graphing Linear Equations
Practice Problems:

1. $m = \dfrac{8}{3}$

2. $m = -\dfrac{5}{3}$

3. $y = 3x + 2$

4. $y = -x + 6$

5. $y = \dfrac{2}{5}x - 3$

6. $y = \dfrac{1}{2}x + 3$

7. $y = \dfrac{3}{2}x + \dfrac{13}{2}$

8.

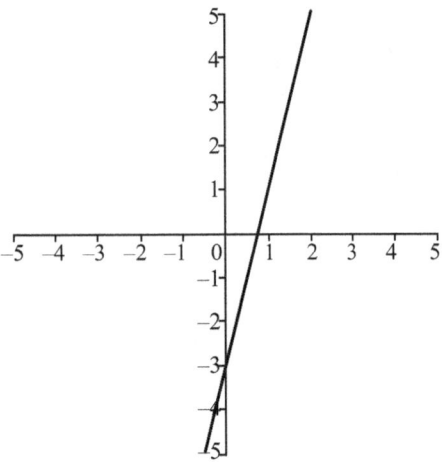

9.

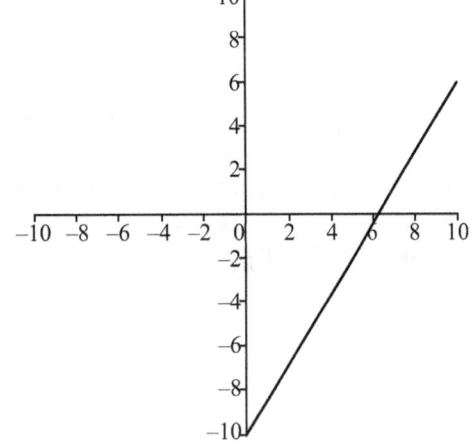

10.

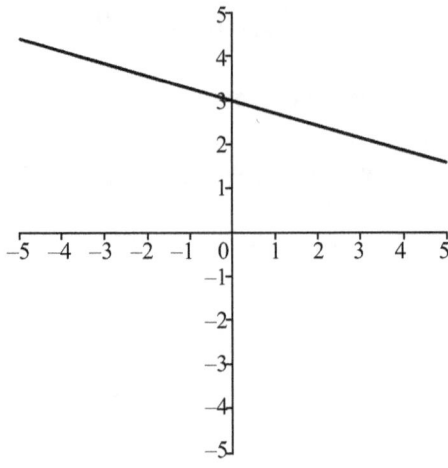

11.

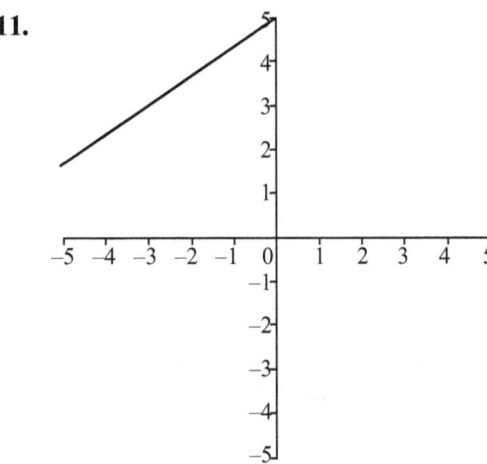

12.

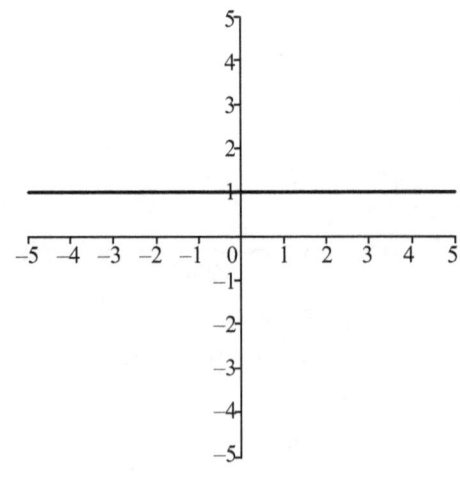

6.3 ACT Problems:

1. **E.** $-\dfrac{2}{5}$ $\dfrac{\text{rise}}{\text{run}} = \dfrac{10 - 8}{-3 - 2} = -\dfrac{2}{5}$

2. **H.** 5   At every point on the y-axis, $x = 0$.
    Plug in $x = 0$. $0 + 2y = 10$ $y = 5$

3. **D.** $y = \dfrac{5}{3}x + 4$ Only answers **A** and **D** have
    a y-intercept of 4. Eliminate the others.

Rearrange the given equation to find the slope. $3x+5y=6 \rightarrow y=-\frac{3}{5}x+\frac{6}{5}$ The slope is $-\frac{3}{5}$, so the perpendicular slope is $\frac{5}{3}$. Choose **D**.

4. **J**. (3,4) Either sketch the line on the graph or plug in the points. $4=\frac{2}{3}(3)+2$ Choose **J**.

5. **E**. $y=-\frac{4}{3}x+4$ Start checking the three easy answers, **A**, **C**, and **E**, because they are in slope-intercept form. The y-intercept is 4. Eliminate **C**. The line is going down from left to right, so we need a negative slope. Eliminate **A**. The slope is $-\frac{4}{3}$. Choose **E**.

If none of those had worked, we would have checked **B** and **D** by plugging in points.

Section 6.4 Graphing Systems of Equations and Inequalities
Practice Problems:
1.
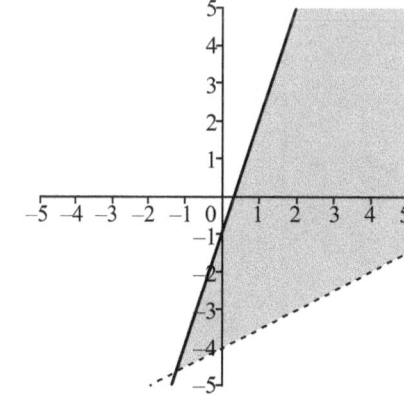

2.
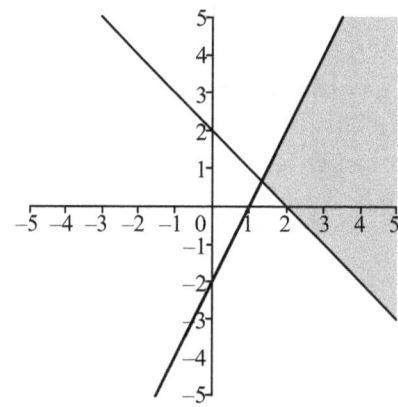

3.
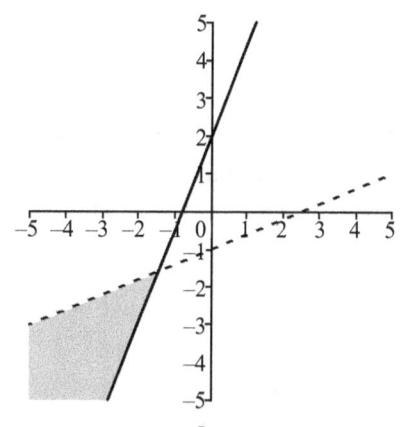

4.
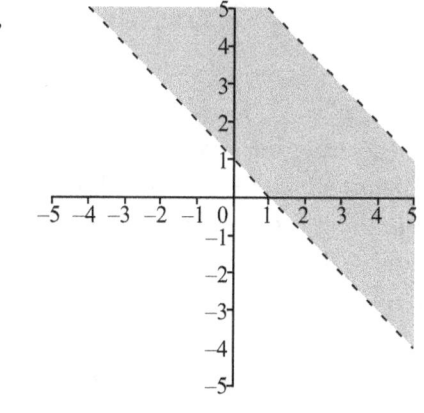

6.4 ACT Problems:
1. **B**. (0,3) Plug in the answers. $4(0)+3(3)>10 \rightarrow 9>10$ That's false. They asked which is NOT a solution. Choose **B**.

2. **F**. $2x+5y\leq10$ First, we have a solid line. All of the answer choices give solid lines. Next, plug in points because the inequalities are not in slope-intercept form. When we plug in points on the line, we want the 2 sides of the inequality to be equal. If we plug in (0,2), we can eliminate **G**, **H** and **J**. If we plug in (5,0), we can also eliminate **K**. The answer must be **F**.

3. **C**. $2x+3y\leq12$ First, we have 2 solid
       $4x-5y\leq20$ lines. Eliminate **E**.
   Next, look at the answers. $2x+3y=12$ Has to be correct because all of the choices have it. Now plug in points. When we plug in points on the line, we want the 2 sides of the inequality to be equal. (0,4) works in $2x+3y=12$, so plug in points from the other line in the second equation. (0,–4) works for **A** and **C**. Eliminate **B** and **D**. To check the greater or less than signs, plug in (0,0). It works in both inequalities in **C**.

145

4. J. $18x-3y=-21$ This is a perfect example of an easy question asked in a confusing way. Two equations have an infinite number of solutions (cross infinitely many times) when they are the same equation. $18x-3y=-21$ Simplifies to $y=6x+7$.

5. A.

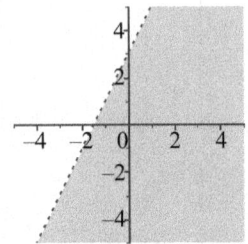

All of the choices have dotted lines, so we check the equation next. $2x-y>-3$ Can be rearranged to $y<2x+3$. Slope is 2 and the y-intercept is 3. Eliminate **B**, **D**, and **E**. Plug in (0,0) to check the shading. $0<2(0)+3$ Is true, so shade (0,0). Pick **A**.

Section 6.5 Graphing Conic Sections Practice Problems:

1.

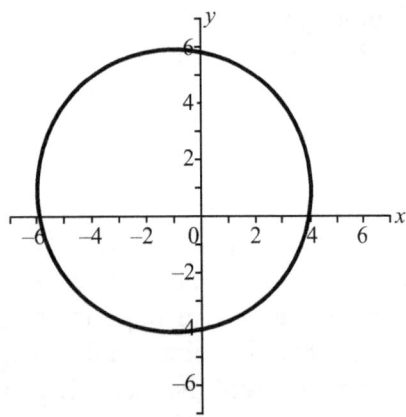

2.

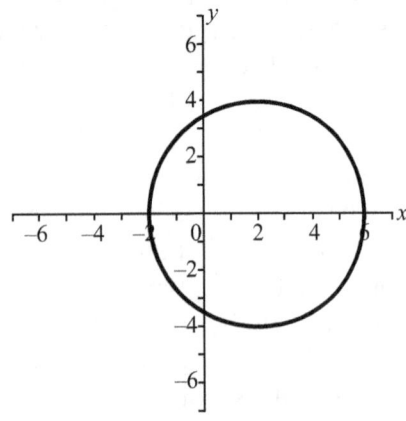

3.

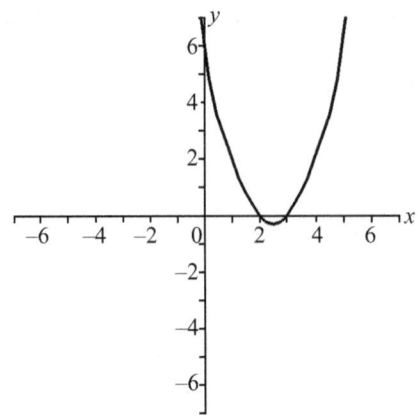

4.

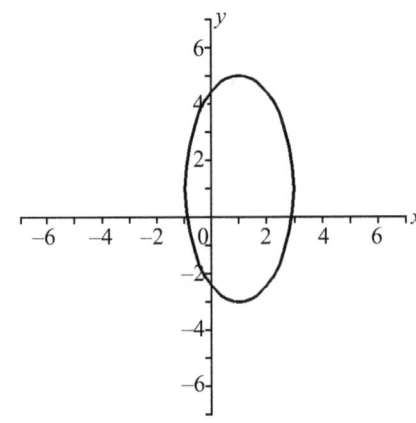

5.

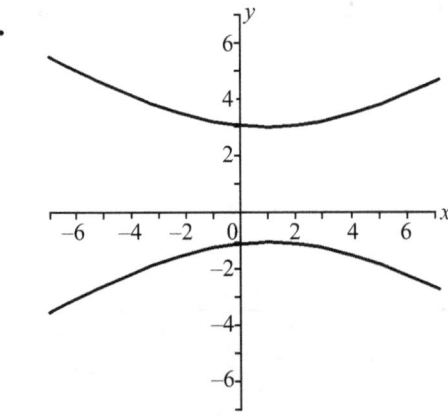

6.5 ACT Problems:

1. E. $\sqrt{13}$ (5,0) The Pythagorean Theorem tells us the radius is $r^2=13$ $\rightarrow$ $r=\sqrt{13}$. Eliminate **A**, **B**, and **C**. Factoring gives us the center of $(x-5)=0$ and $y=0$, so the center is (5,0). Choose **E**.

146

2. **K.** I, II, and III   The question tells us that c is the height at time $t = 0$. Try sketching another curve that only changes the starting height.

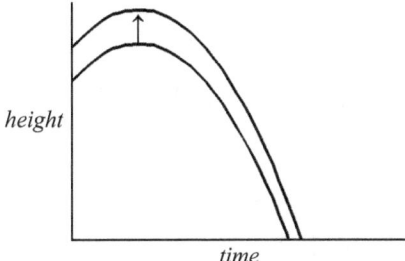

You can see that the h-intercept, t-intercept, and maximum height all change. Choose **K**.

3. **A.** $x^2 + 2y^2 = 16$ Eliminate answers that you know are wrong. **C** is the standard form of a circle, and **E** is a parabola. We know an ellipse has a plus sign between the x and y, so eliminate **B**. **D** can be reduced by 4 to give $x^2 + y^2 = 2$, which is a circle. The only one left is **A**. If you think of an ellipse as an uneven circle, answer **A** makes more sense.

4. **H.** $(x+4)^2 + (y-3)^2 = 25$ The center is at $(-4, 3)$. Remember to use the opposite signs in the parentheses. The answer must be **H**.

5. **A.**   This equation is a mess. If you remember completing the square, you can put it in standard form, but it's not worth using that much time. Try to eliminate answers. First, the x and y are both squared, so it can't be a parabola. Eliminate **C**. The x^2 and y^2 do not have the same coefficients (9 and -4), so it can't be a circle. Eliminate **E**. Finally, the x^2 and y^2 terms are subtracted, so it must be a hyperbola. Choose **A**. If all else fails, plug in points. Just don't use the centers.

Section 6.6 Transformations
Practice Problems:
1. Shift right 3 units and down 4 units.
2. Rotate 90° clockwise around the origin.
3. Reflect across the y axis.

4.

5.

6.

6.6 ACT Problems:
1. **B.** $x = 0$ Draw the mirror.

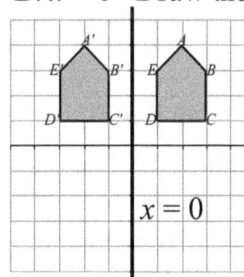

2. **G.** (3,2)   Move the square 2 units right and 4 units down. All of the answers have an x coordinate of 3. The two vertices with an x coordinate of 3 are (3,2) and (3,-3). Only one is a choice.

147

3. A.

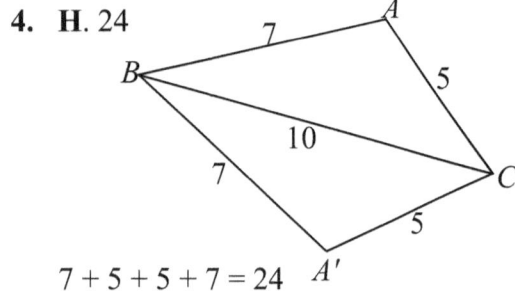

To flip across the x axis, flip the graph down.

4. H. 24

$7 + 5 + 5 + 7 = 24$

Coordinate Geometry Review

1. B. $y = x + 5$ To find the equation, we need the slope and the y-intercept. The y-intercept is easier. The line crosses y at 5. The equation is $y = mx + 5$. Eliminate **D** and **E**. Now find the slope.

$$m = \frac{5-0}{0-(-5)} = \frac{5}{5} = 1$$ Choose **B**.

2. F. (4.5, 4)   Average the x's and the y's.

$$\frac{1+8}{2} = 4.5 \text{ and } \frac{10-2}{2} = 4$$ Choose **F**.

3. C. $y = \frac{2}{3}x + 2$ First we need to put the equation in slope-intercept form.

$$2y + 3x = 8 \rightarrow y = -\frac{3}{2}x + 4$$ This line has a

slope of $-\frac{3}{2}$, so the perpendicular slope is

$\frac{2}{3}$. Eliminate **A**, **B**, and **D**. Plug in the point (3,4) to **C** and **E**. **C** works.

$$4 = \frac{2}{3}(3) + 2$$

4. K.

The inequality has an "or equal" sign, so we need a closed circle. Eliminate **F**, **G**, and **H**. Solve the inequality. $x \geq -2$ Choose **K**.

5. C. $5\sqrt{2}$ To find distance, use the Pythagorean Theorem. The x-distance is $-1 - (-6) = 5$. The y-distance is $4 - (-1) = 5$. $5^2 + 5^2 = c^2$ $c = \sqrt{50} = 5\sqrt{2}$

6. F. $\frac{5}{2}$ Slope is rise over run. Rise is y. To get from 8 to 3, go down 5. Rise = –5. Run is x, and –3 to –5 is left 2, or –2.

$$\frac{rise}{run} = \frac{-5}{-2} = \frac{5}{2}.$$

7. B. a parabola   The x is squared, but the y is not. It must be a parabola.

8. F.

The pool is being filled, so the line must go up. Eliminate **J** and **K**. When we add another hose, the water will rise faster. Graph **F** shows the water start rising faster.

9. B. (–1, 8), (4, 7), (5, 3), (–2, 2)   2 units to the right means add 2 to each x coordinate and 3 units up means add 3 to each y coordinate.

10. F.

The easiest way to find the answer is to use the x- and y-intercepts. If $x = 0$, there are no student tickets sold. How many adult tickets must be sold? $5y = 300 \rightarrow y = 60$. On these graphs, you can easily estimate. Eliminate **G**, **J**, and **K** because the lines don't cross the y-axis anywhere near 60. Now find the x-intercept. If there are no adult tickets sold, then $3x = 300 \rightarrow x = 100$. Choose **F**.

Section 7.1
Basics of Plane Geometry

Vocabulary

point – a position in space, it has no length or size (point A)   • A

line – a continuous straight path, it extends infinitely in both directions   ($\overleftrightarrow{AB}$)

$\longleftrightarrow$
$A \qquad\qquad B$

line segment – part of a line with two endpoints, the length can be measured   ($\overline{AB}$)

$A \qquad\qquad B$

ray – part of a line with one endpoint, it extends infinitely in one direction   ($\overrightarrow{AB}$)

$A \qquad\qquad B$

parallel – lines or line segments that never intersect   ($\overline{AB} \parallel \overline{CD}$)

$A \qquad\qquad B$
$C \qquad\qquad D$

perpendicular – lines or line segments that intersect to form right angles   ($\overline{AB} \perp \overline{CD}$)

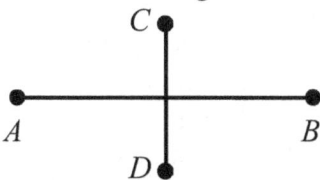

skew – lines or line segments in three dimensional space
that do not intersect and are not parallel   ($\overline{AB}$ and $\overline{CD}$ are skew segments)

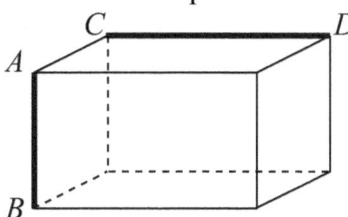

plane – a flat surface extending infinitely in all directions

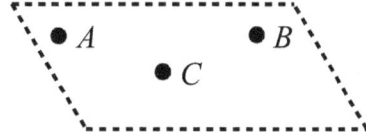

Example 1

Opposite walls in a hallway are examples of what?

    A. parallel lines
    B. perpendicular lines
    C. parallel planes
    D. perpendicular planes
    E. skew planes

First, walls are planes, not lines, so eliminate **A** and **B**. Opposite walls do not cross, so eliminate **D**. We can eliminate **E** because there are no such things as skew planes. Planes extend forever, so if two planes are not parallel, they will eventually cross. Answer **C** is the only one that makes sense.

7.1 Practice Problems: Answers on p. 183

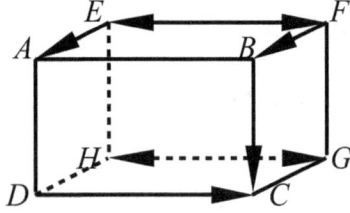

    1. Name all the points
    2. Name all the lines
    3. Name all the segments
    4. Name all the rays (pay attention to endpoints)
    5. Name a pair of parallel lines
    6. Name a pair of perpendicular segments
    7. Name a plane

7.1 ACT Problems: Answers on p. 183

    1. If line $\overleftrightarrow{AB}$ and line $\overleftrightarrow{CD}$ are both perpendicular to line $\overleftrightarrow{YZ}$, which of the following statements must be true? Assume all lines are in the same plane.
        A. $\overleftrightarrow{AB} \perp \overleftrightarrow{CD}$
        B. $\overleftrightarrow{AB} \| \overleftrightarrow{CD}$
        C. $\overrightarrow{AB}$ and $\overleftrightarrow{CD}$ are skew
        D. $\overline{AB}$, $\overline{CD}$, and $\overline{YZ}$ form a triangle
        E. $\overleftrightarrow{AB}$ and $\overleftrightarrow{CD}$ intersect at point Z

    2. Which of the following figures in a plane separates it into half-planes?
        F. A line
        G. A ray
        H. An angle
        J. A point
        K. A line segment

    3. The notation AB refers to which of the following?
        A. point A and point B
        B. the length of line $\overleftrightarrow{AB}$
        C. the length of ray $\overrightarrow{AB}$
        D. the length of segment $\overline{AB}$
        E. the measure of angle AB

Section 7.2
Angles

An **angle** is made from two rays with a common endpoint called a **vertex**.

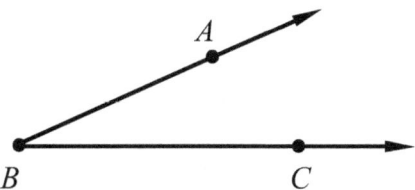

Point *B* is the vertex. The angle can be named ∠*ABC*, ∠*CBA*, or ∠*B*. The measure of this angle is 40°. This is written *m*∠*B* = 40°.

An **acute** angle measures less than 90°.

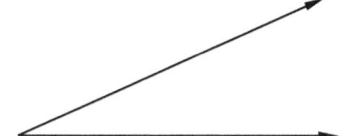

A **right** angle measures exactly 90°.

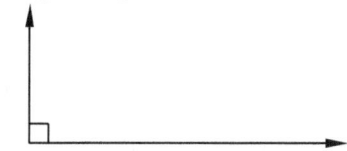

An **obtuse** angle measures more than 90°.

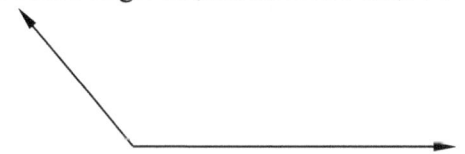

A **straight** angle measures exactly 180°.

Congruent – angles that have the same measure.

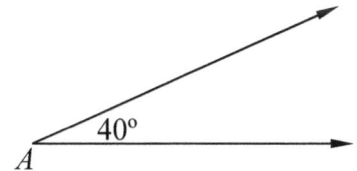

∠*A* is congruent to ∠*B*. ∠*A* ≅ ∠*B*

Complementary – angles that add up to 90°.

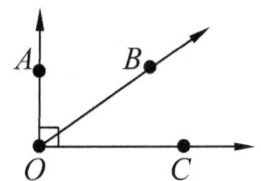

∠*AOB* and ∠*BOC* are **complementary** angles.

Supplementary – angles that add up to 180°.

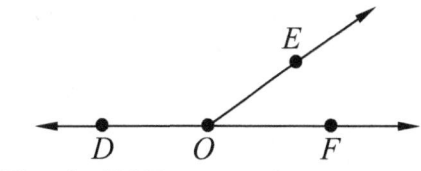

∠*DOE* and ∠*EOF* are **supplementary** angles.

Parallel lines cut by a transversal

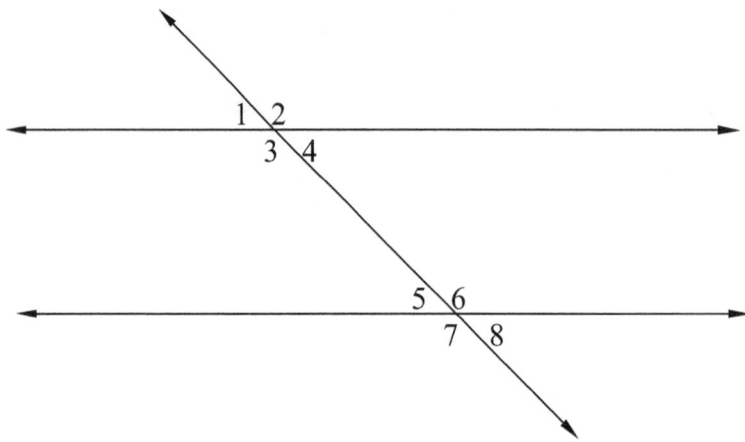

Transversal – a line that crosses parallel lines

Adjacent – angles that touch on a shared side. For example, ∠1 is adjacent to ∠2 and ∠3.

Vertical – angles that are across from each other when two lines cross. For example, ∠1 and ∠4 are vertical angles. Vertical angles are congruent.

Corresponding – angles that are in the same position on two parallel lines. One angle can slide over to sit exactly on top of the other without rotating. For example, ∠1 and ∠5 are corresponding. Corresponding angles are congruent.

Alternate – the two angles are on opposite sides of the transversal.

Same-side – the two angles are on the same side of the transversal.

Interior – the angles are inside of the parallel lines. (∠3, ∠4, ∠5, ∠6)

Exterior – the angles are outside the parallel lines. (∠1, ∠2, ∠7, ∠8)

Congruent and Supplementary pairs

Alternate interior angles (∠3 and ∠6) and **alternate exterior** angles (∠1 and ∠8) are congruent.
Same-side interior angles (∠3 and ∠5) and **same-side exterior** angles (∠1 and ∠7) are supplementary.

Congruent and supplementary pairs are really easy to figure out if you pay attention to acute and obtuse angles. Any time you have parallel lines cut by a single transversal, all of the acute angles are equal to each other. All of the obtuse angles are equal to each other. Any acute plus any obtuse adds to 180°.

Example 1

In the figure to the right, parallel lines r and s are intersected by line t. What is the measure of angle α?

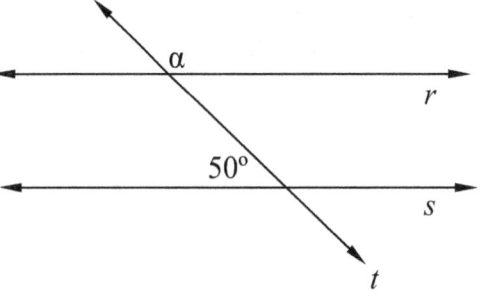

Angle α is obtuse, so we know it is more than 90°. We're given an acute angle of 50°. Since we have parallel lines and a transversal, the acute and obtuse angles must be supplementary. $50 + \alpha = 180$.

$\alpha = 130°$

Example 2

In the figure to the right, line l is parallel to line m. Transversals t and u intersect at point A on l and intersect m at points C and B, respectively. Point X is on m, the measure of $\angle ACX$ is 140°, and the measure of $\angle BAC$ is 80°. What is the measure of $\angle ABC$?

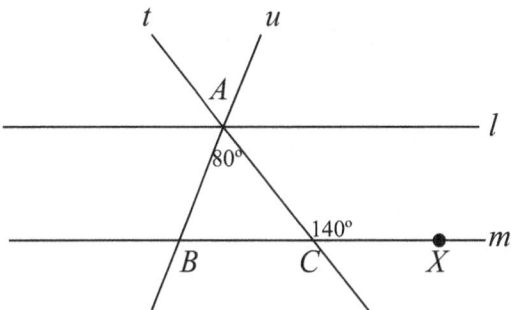

This question is way too long. We could get all of that information by looking at the picture. Start labeling angles you know; ignore line u and label all obtuse angles 140° and all acute angles $180 - 140 = 40°$ for line t only.

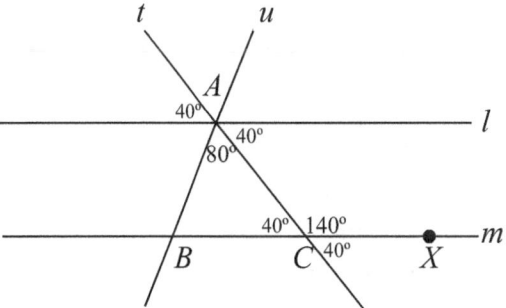

What else do we know? 40°, 80°, and an unlabeled acute angle on line u make a straight line, so they add up to 180°. Also, 40°, 80°, and $\angle ABC$ make a triangle, so they add up to 180°. Either way you see it, the acute angles on line u all equal $180 - 40 - 80 = 60°$

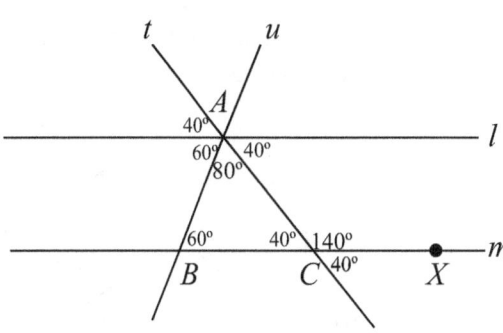

$\angle ABC = 60°$

153

7.2 Practice Problems: Answers on p. 183
In the figure below, find the following.

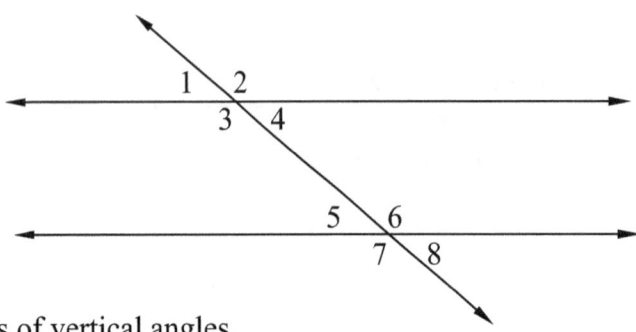

1. All pairs of vertical angles.
2. All pairs of corresponding angles.
3. All pairs of alternate interior angles.
4. All pairs of alternate exterior angles.
5. All pairs of same-side interior angles.
6. All pairs of same-side exterior angles.
7. Which of the above pairs are congruent? Supplementary? Neither?

7.2 ACT Problems: Answers on p. 183

1. In the figure below, m∠3 = 140°, $\overleftrightarrow{AB} \parallel \overleftrightarrow{CD}$, and $\overleftrightarrow{AB}$ and $\overleftrightarrow{CD}$ are cut by a transversal, *t*. What is the product of m∠2 and m∠5?

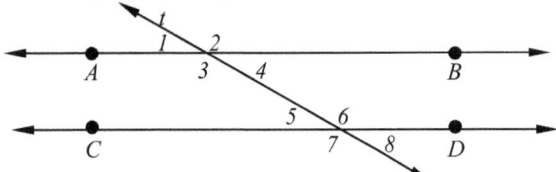

A. 40°
B. 140°
C. 180°
D. 560°
E. 5600°

2. In the figure below, l_1 is parallel to l_2, l_3 is parallel to l_4, and the lines intersect as shown. What is the measure of angle *z*?

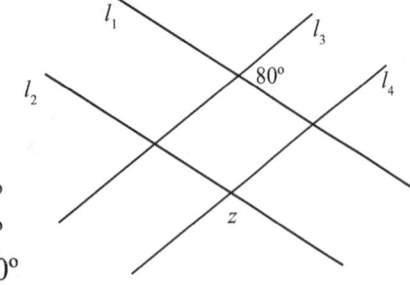

F. 80°
G. 90°
H. 100°
J. 120°
K. Cannot be determined from the given information

3. In the figure below, *B* is on $\overline{DE}$ and $\overline{DE} \parallel \overline{AC}$. Which of the following angle congruences must hold?

A. ∠1 ≅ ∠2
B. ∠1 ≅ ∠4
C. ∠2 ≅ ∠3
D. ∠2 ≅ ∠4
E. ∠3 ≅ ∠4

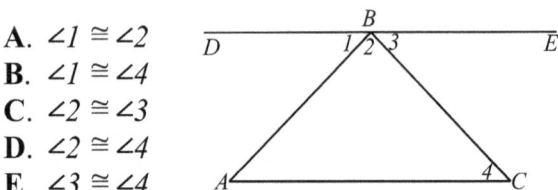

4. In the figure below, line *l* is parallel to line *m*. Transversals *t* and *u* intersect at point *A* on *l* and intersect *m* at points *C* and *B*, respectively. Point *X* is on *m*, the measure of ∠ACX is 130°, and the measure of ∠BAC is 80°. How many of the angles formed by rays of *l*, *m*, *t*, and *u* have measures of 50°?

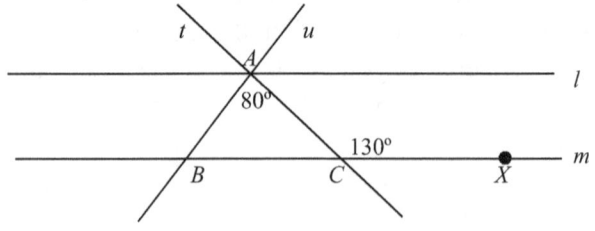

F. 4
G. 6
H. 8
J. 10
K. 12

154

5. In the figure below, parallel lines *a* and *b* intersect parallel lines *c* and *d*. If it can be determined, what is the sum of the degree measures of ∠1 and ∠2?

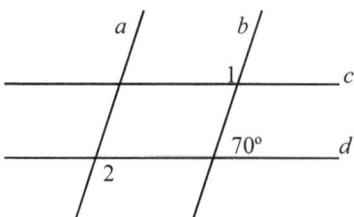

A. 220°
B. 180°
C. 140°
D. 110°
E. Cannot be determined from the given information

Section 7.3
Quadrilaterals and Polygons

A **quadrilateral** is a shape with four sides.

A **polygon** is a shape with many (three or more) sides.

A **parallelogram** is a quadrilateral with *2 pairs of congruent parallel lines*.

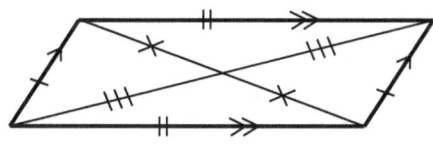
Both pairs of opposite angles are congruent (the two acute angles are equal and the two obtuse angles are equal). Consecutive angles are supplementary (one acute plus one obtuse equals 180°). The diagonals bisect (each diagonal cuts the other in half).

A **rectangle** is a parallelogram with *right angles*.

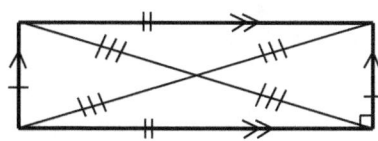
All of the properties of a parallelogram are true for a rectangle. All four angles are congruent (all are 90°). The diagonals are congruent.

A **rhombus** is a parallelogram with *congruent sides*.

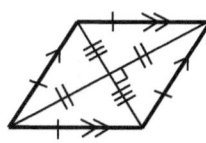
All of the properties of a parallelogram are true for a rhombus. The diagonals are perpendicular.

155

A **square** is a rhombus and a rectangle.

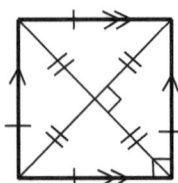

All of the properties of a parallelogram, rectangle, and rhombus are true.

A **trapezoid** has *only one* pair of parallel sides.

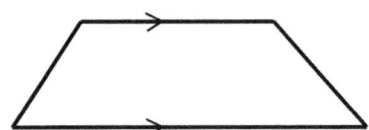

One pair of sides is parallel.
Nothing has to be congruent.

Other polygons

- A **regular** polygon has all congruent sides and congruent angles.

- A **triangle** has 3 sides.
- A **quadrilateral** has 4 sides.
- A **pentagon** has 5 sides.
- A **hexagon** has 6 sides.
- A **heptagon** has 7 sides.
- An **octagon** has 8 sides.
- A **nonagon** has 9 sides.
- A **decagon** has 10 sides.

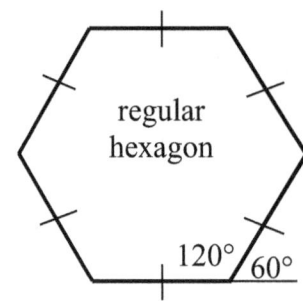

regular hexagon

$120°$ $60°$

- The sum of the interior angles is $180(n-2)°$ where n is the number of sides.
- The sum of the exterior angles is $360°$ regardless of the number of sides.

- A **concave** polynomial bends back in on itself (makes a cave).
- A **convex** polynomial does not bend back in on itself.

concave pentagon

convex pentagon

Example 1
What is the sum of the interior angles of a heptagon?

Method 1
If you remember the formula, plug in numbers.
A heptagon has 7 sides, so $n = 7$.

$180(n - 2)°$
$180(7 - 2)$
$180(5) = 900°$

Method 2
If you don't remember the formula, remember that the sum of the exterior angles is always 360°.
Assume all of the angles are congruent.
Each exterior angle would be $360 \div 7 \approx 51.4$

The interior angle would have to be $180 - 51.4 = 128.6$.

If all of the angles are the same, then the sum would be
$128.7(7) \approx 900$

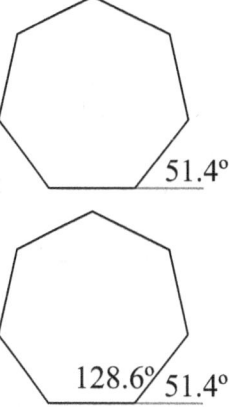

Example 2
Which of the following statements is false?
- **A**. A rectangle is a parallelogram
- **B**. A square is a rectangle
- **C**. A rhombus is a square
- **D**. A square is a parallelogram
- **E**. A rectangle is a convex polygon

The correct answer is **C**. A square is a rhombus, but a rhombus does not have to be a square. All of the other statements are true.

7.3 Practice Problems: Answers on p. 183

1. What is the name for a quadrilateral with bisecting diagonals?

2. What is the name for a parallelogram with congruent diagonals?

3. In parallelogram $ABCD$, $\angle A = 40°$. What are the measures of the other three angles?

4. What is the sum of the interior angles of a pentagon?

5. What is the sum of the exterior angles of a nonagon?

6. What is the name for a quadrilateral where one pair of sides is parallel but not congruent?

7. What is the measure of an interior angle of a regular octagon?

1. The perimeter of a figure is 4 times the length of one side. This figure could be:
   A. A square
   B. A rhombus
   C. A trapezoid
   D. All of the above
   E. None of the above

2. In any parallelogram *ABCD*, it is always true that the measures of ∠*ABC* and ∠*BCD*:
   F. add up to 180°
   G. add up to 90°
   H. are each greater than 90°
   J. are each 90°
   K. are each less than 90°

3. In the figure below, which is drawn to scale, exactly one of the following subfigures is a parallelogram. Which one?

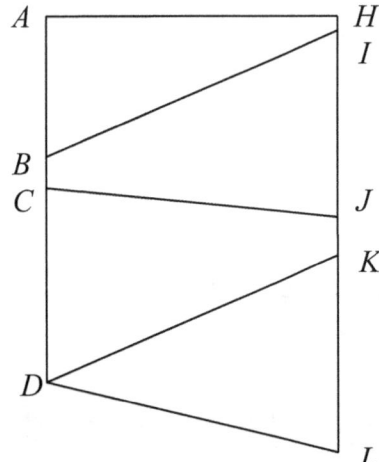

   A. *ADLH*
   B. *ACJH*
   C. *BDKI*
   D. *CJKD*
   E. *CJLD*

4. In pentagon *PENTA*, shown below, ∠*P* measures 40°. What is the total measure of the other four angles?

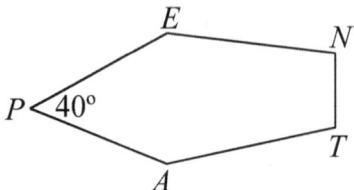

   F. 140°
   G. 210°
   H. 320°
   J. 360°
   K. 500°

5. In the figure below, 2 nonadjacent sides of a regular pentagon (5 congruent sides and 5 congruent interior angles) are extended until they meet at point *X*. What is the measure of ∠*X*?

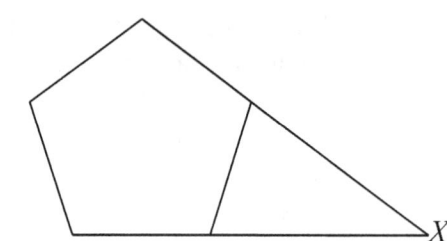

   A. 18°
   B. 30°
   C. 36°
   D. 45°
   E. 72°

Section 7.4
Triangles

- In any triangle, the measures of the angles add up to **180°**.
- If two sides of a triangle are congruent, then the angles opposite those sides are congruent.
- The sum of the lengths of any two sides of a triangle is greater than the length of the third side. (When you add the two short sides, the sum has to be longer than the long side.)
- Triangles can be classified by congruent sides and angles, or they can be classified by angle measures.

An **equilateral triangle** has 3 congruent sides and 3 congruent angles.

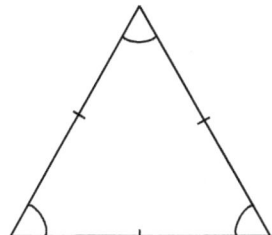

An **isosceles triangle** has 2 congruent sides and 2 congruent angles (called base angles). The third angle is called the vertex angle.

A **scalene triangle** has three sides with different lengths and three angles with different measures.

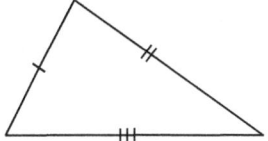

A **right triangle** has one 90° angle.

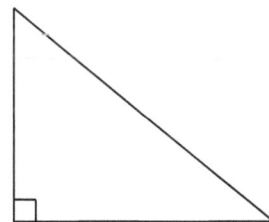

An **acute triangle** has all acute angles.

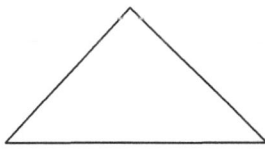

An **obtuse triangle** has one obtuse angle.

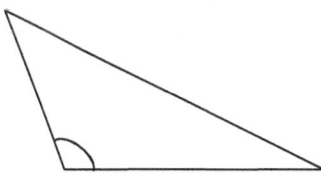

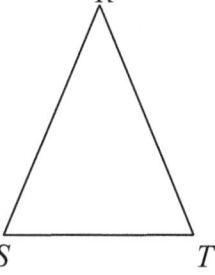

Example 1

In $\triangle RST$, $\overline{RS} \cong \overline{RT}$, and the measure of $\angle R$ is 46°. What is the measure of $\angle S$?

We know this is an isosceles triangle because two sides are congruent.
If two sides are congruent, then the two angles across from them are congruent, therefore $\angle S \cong \angle T$. We also know that the three angles add up to 180°.

$$46 + x + x = 180$$
$$x = 67$$

$\angle S$ and $\angle T$ are each equal to 67°.

Example 2

In the figure to the right, ∠*BAC* measures 30°, ∠*ABC* measures 110°, and points *B*, *C*, and *D* are collinear. What is the measure of ∠*ACD*?

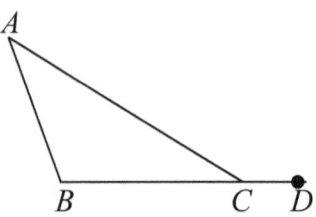

If we know two angles in a triangle, we can easily find the third.

$$30+110+x=180$$
$$x=40$$

m∠*ACB* = 40, and we know that ∠*ACB* and ∠*ACD* are supplementary.

$$40+y=180$$
$$y=140$$

The measure of ∠*ACD* is 140°.

7.4 Practice Problems: Answers on p. 184

1. In an equilateral triangle, what are the measures of each angle?
2. One of the angles in an isosceles triangle is 25°. What are the possible measurements of the other two angles?
3. Find the measure of ∠*x*.

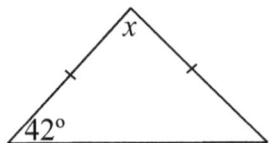

4. Find the measure of ∠*x*.

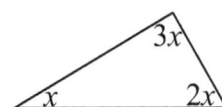

Are the measurements of these triangles possible? Explain.

5.

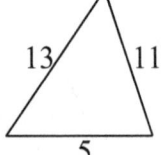

6.

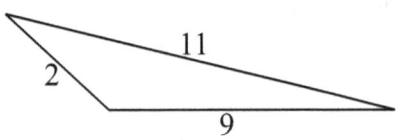

160

1. In the figure below, A, D, B, and G are collinear. If $\angle CAD$ measures $81°$, $\angle BCD$ measures $33°$, and $\angle CBG$ measures $130°$, what is the degree measure of $\angle ACD$?

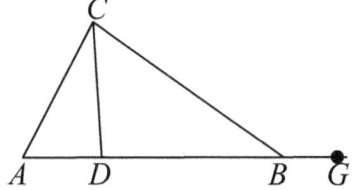

   A. $11°$
   B. $14°$
   C. $16°$
   D. $49°$
   E. $97°$

2. Each of 3 lines crosses the other 2 lines, as shown below. Which of the following relationships, involving angle measures (in degrees) *must* be true?

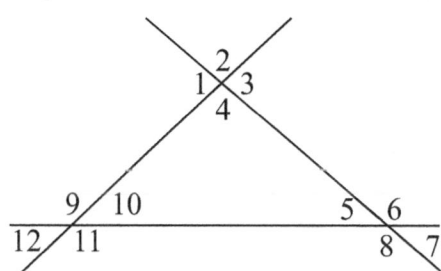

   I. $m\angle 4 + m\angle 5 + m\angle 10 = 180°$
   II. $m\angle 2 + m\angle 7 + m\angle 12 = 180°$
   III. $m\angle 2 + m\angle 7 + m\angle 10 = 180°$

   F. I only
   G. II only
   H. III only
   J. I and II only
   K. I, II, and III

3. Which of the following choices could NOT be the lengths of the sides of a triangle?
   A. 5, 5, 5
   B. 2, 3, 6
   C. 4, 8, 8
   D. 3, 4, 6
   E. 6, 8, 10

4. For all triangles $\triangle XYZ$ where side $\overline{XZ}$ is longer than side $\overline{YZ}$, such as the triangle shown below, which of the following statements is true?

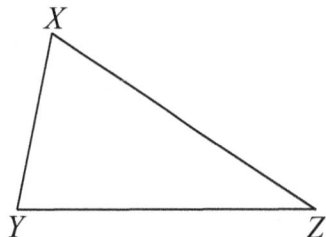

   F. The measure of $\angle X$ is always less than the measure of $\angle Y$.
   G. The measure of $\angle X$ is always equal to the measure of $\angle Y$.
   H. The measure of $\angle X$ is always greater than the measure of $\angle Y$.
   J. The measure of $\angle X$ is sometimes less than the measure of $\angle Y$ and sometimes equal to the measure of $\angle Y$.
   K. The measure of $\angle X$ is sometimes greater than the measure of $\angle Y$ and sometimes equal to the measure of $\angle Y$.

5. In the standard (x,y) coordinate plane below, $\triangle ABC$ and $\triangle CDE$ are isosceles right triangles with equal measurements. Points A, B, C, and D are located on the axes as shown. Which of the following could be the coordinates of point E?

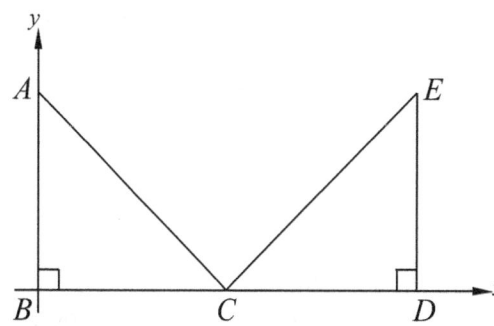

   A. $(0,10)$
   B. $(5, 0)$
   C. $(5,10)$
   D. $(10,0)$
   E. $(10,5)$

Section 7.5
Right Triangles

A right triangle is made up of two *legs* (*a* and *b*) that meet at a right angle. The longer third side is called the *hypotenuse* (*c*).

The **Pythagorean Theorem** states the relationship between the sides: $a^2 + b^2 = c^2$

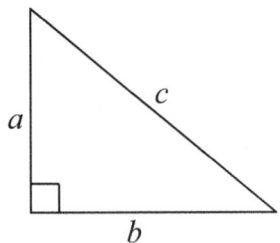

Common **Pythagorean triples** (whole numbers that satisfy the Pythagorean Theorem):

$3 - 4 - 5$	$5 - 12 - 13$
$6 - 8 - 10$	$8 - 15 - 17$
$9 - 12 - 15$	$7 - 24 - 25$

Special right triangles

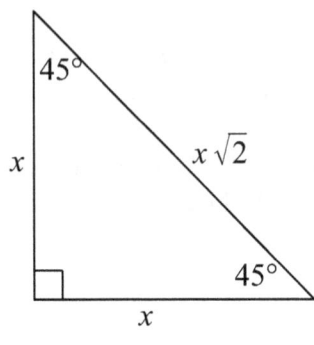

$45 - 45 - 90$

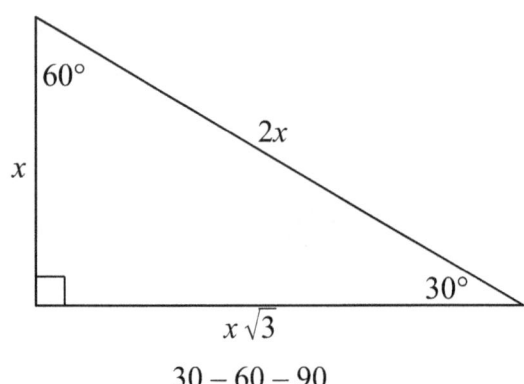

$30 - 60 - 90$

Example 1
Solve for *x*:

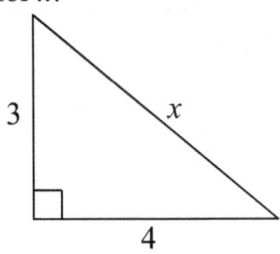

Use the Pythagorean Theorem.
Plug in 3 and 4 for *a* and *b*. They are both legs, so it doesn't matter which one is *a* and which one is *b*, but *x* is the hypotenuse, so it has to be *c*.
Don't forget to take the square root to solve.

$$a^2 + b^2 = c^2$$
$$3^2 + 4^2 = x^2$$
$$9 + 16 = x^2$$
$$25 = x^2$$
$$x = 5$$

Example 2

Δ*XYZ* is an isosceles right triangle. The measure of side $\overline{XZ}$ is 8. Find the measure of side $\overline{XY}$.

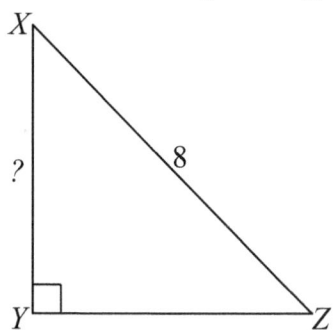

Method 1 – remember 45 – 45 – 90 triangles

This is a special right triangle, so we know that when side $\overline{XY}$ is *x*, side $\overline{XZ}$ is $x\sqrt{2}$. Set up an equation.

$$x\sqrt{2}=8$$

Solve.

$$x=\frac{8}{\sqrt{2}}$$

Rationalize.

$$x=\frac{8}{\sqrt{2}}\times\frac{\sqrt{2}}{\sqrt{2}}=\frac{8\sqrt{2}}{2}=4\sqrt{2}$$

$$\overline{XY}=4\sqrt{2}$$

Method 2 – Pythagorean Theorem

This is an isosceles right triangle, so we know that the legs are equal.
Set up the Pythagorean Theorem plugging in *x* for both legs.
Solve.

$$x^2+x^2=8^2$$
$$2x^2=64$$
$$x^2=32$$
$$x=\sqrt{32}$$

Simplify.

$$x=4\sqrt{2}$$

7.5 Practice Problems: Answers on p. 184

Solve for *x*.

1.

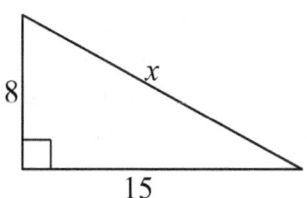

2.

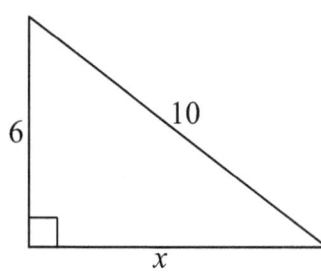

3.

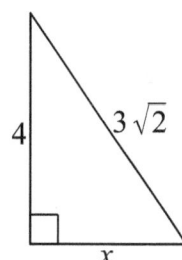

4.

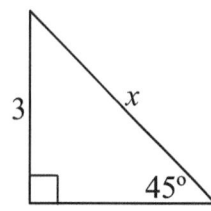

163

5.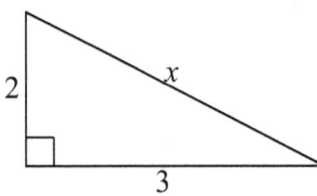

6.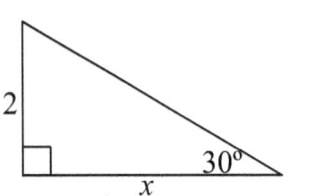

7.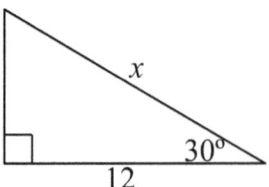

7.5 ACT Problems: Answers on p. 184-185

1. How many centimeters long is the diagonal of a rectangle that is 7 centimeters wide and 12 centimeters long, as shown below?

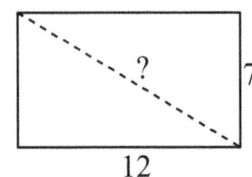

A. 15
B. 19
C. $\sqrt{84}$
D. $\sqrt{93}$
E. $\sqrt{193}$

2. In the isosceles right triangle below, $XY = 6$ feet. What is the sum of the lengths of $\overline{YZ}$ and $\overline{XZ}$?

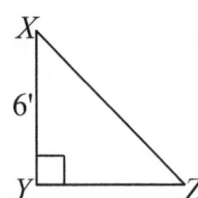

F. 6
G. $6\sqrt{2}$
H. $12+\sqrt{2}$
J. $6+6\sqrt{2}$
K. $12\sqrt{2}$

3. What is the measure of $\angle ACB$ in the triangle below?

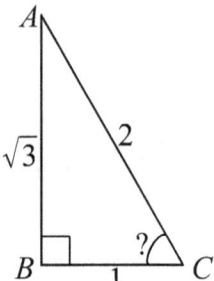

A. 15°
B. 30°
C. 45°
D. 60°
E. 75°

4. The base of a ladder is placed 8 feet from a house, and the top of the ladder touches the house 16 feet above the ground. If the house creates a 90° angle with the ground, how long is the ladder to the nearest tenth of a foot?

F. 13.9 feet
G. 17.9 feet
H. 16.0 feet
J. 24.0 feet
K. 32.0 feet

5. Given the information in the diagram below, including the fact that $\triangle ABC$ is isosceles, how many feet long is $\overline{CD}$?

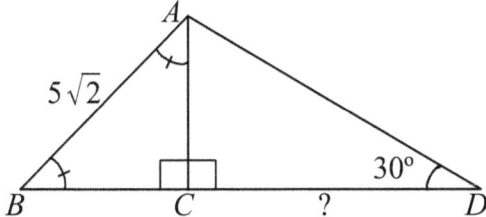

- **A.** 5
- **B.** $5\sqrt{2}$
- **C.** $5\sqrt{3}$
- **D.** $5\sqrt{6}$
- **E.** 10

Section 7.6
Similar Triangles

Similar triangles (or other figures) are proportional. They have the same shape but not the same size.

$\triangle ABC \sim \triangle DEF$ means triangle ABC is similar to triangle DEF.
- $\angle A$ is congruent to $\angle D$; $\angle B$ is congruent to $\angle E$; $\angle C$ is congruent to $\angle F$
- $\overline{AB}$ corresponds to $\overline{DE}$; $\overline{BC}$ corresponds to $\overline{EF}$; $\overline{AC}$ corresponds to $\overline{DF}$
- The lengths of the three corresponding pairs of sides are proportional.

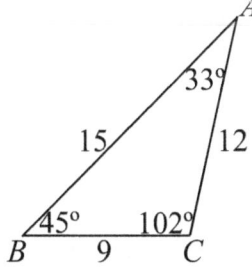
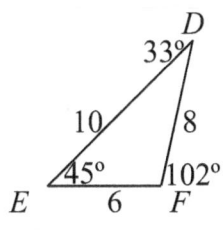

$$\frac{AB}{DE}=\frac{BC}{EF}=\frac{AC}{DF}=\frac{3}{2}$$

Example 1
Given that $\triangle TRI \sim \triangle ANG$, find the length of x.

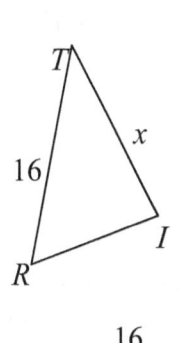
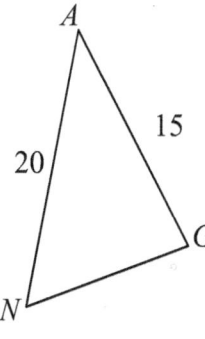

First, find a pair of sides that match up and that you know both measures: 16 and 20

Make a fraction out of them.

$\dfrac{16}{20}$

Next, the x lines up with the 15. Whichever triangle you started with on top, keep on top.

$\dfrac{16}{20}=\dfrac{x}{15}$

Solve by cross-multiplying.

$240=20x$
$12=x$

165

Example 2

$\triangle CAT \sim \triangle DOG$. Find x, y, a, and b.

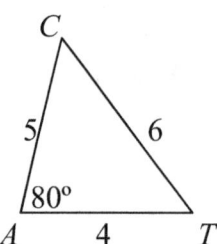
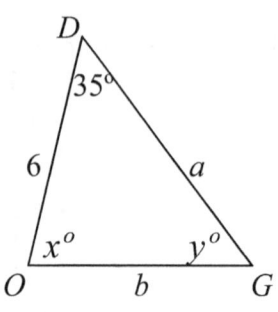

Start with the easiest one. $\angle A$ corresponds with $\angle O$, so $x = 80°$

Next, we can find y because the angles of a triangle total 180°.

$$35° + 80° + y = 180°$$
$$y = 65°$$

Find a and b using ratios.

$$\frac{5}{6} = \frac{6}{a}$$

Cross-multiply.

$$5a = 36$$

Solve.

$$a = 7.2$$

Find b the same way.

$$\frac{5}{6} = \frac{4}{b}$$
$$5b = 24$$
$$b = 4.8$$

7.6 Practice Problems: Answers on p. 185

$\triangle ABC \sim \triangle XYZ$ Find the missing sides and angles.

1.

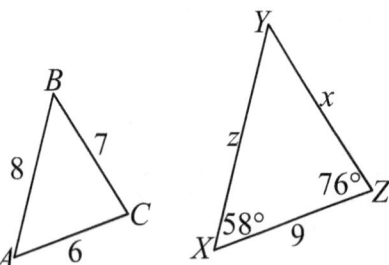

2.

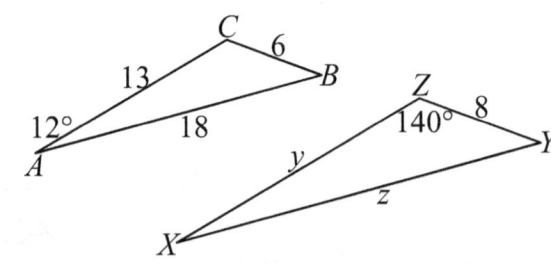

3.

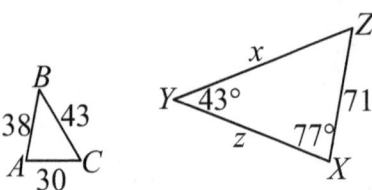

4. $\overline{BC} \| \overline{YZ}$

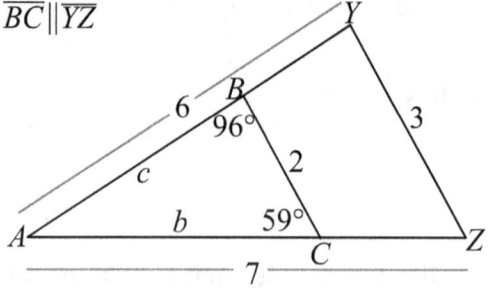

5. If the lengths of the sides of one triangle are 8 inches, 10 inches, and 12 inches, respectively, what is the perimeter, in inches, of a similar triangle whose longest side is 4 inches?

1. The sides of one triangle are 16 inches, 17 inches, and 20 inches long, respectively. In a second triangle similar to the first, the shortest side is 9 inches long. To the nearest tenth of an inch, what is the length of the longest side of the second triangle?
   A. 7.2 in.
   B. 9.8 in.
   C. 10.6 in.
   D. 11.3 in.
   E. 18.8 in.

2. In the figure below, P lies on $\overline{AB}$, Q lies on $\overline{AC}$, and $\overline{PQ}$ is parallel to $\overline{BC}$. If $\overline{PQ}$ is 5 units long, $\overline{AQ}$ is 3 units long, and $\overline{BC}$ is 15 units long, how many units long is $\overline{CQ}$?

   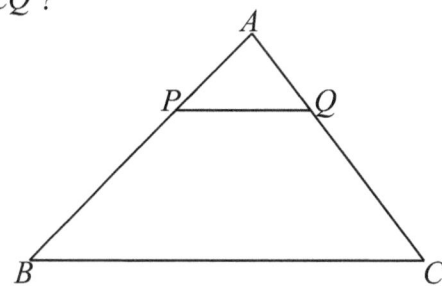

   F. 5
   G. 6
   H. 9
   J. 12
   K. 15

3. The figure below shows 2 triangles, where $\triangle ABC \sim \triangle A'B'C'$. In these similar triangles, $a = 9$, $b = 12$, $c = 15$, and $a' = 15$, with all dimensions given in feet. What is the value of b'?

   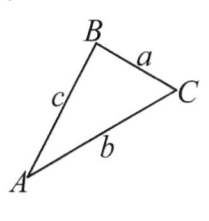

   A. 14 feet
   B. 16 feet
   C. 18 feet
   D. 20 feet
   E. 22 feet

4. For $\triangle ABC$ below, D and E are points on the sides of the triangle. If $\overline{AB}$ is parallel to $\overline{DE}$, what is the measure of $\angle ACE$?

   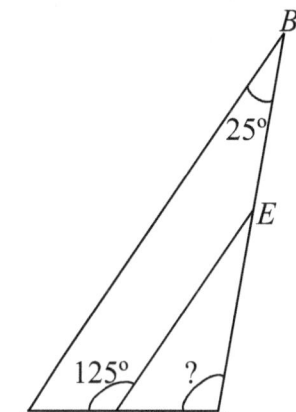

   F. 80°
   G. 100°
   H. 125°
   J. 150°
   K. 155°

5. Bobbi's design for a rectangular window is shown below. It is made up of seven triangles. The two white triangles will be yellow and are congruent ($\triangle ABC \cong \triangle AED$). The other five triangles will be red. The measures of $\angle BAC$, $\angle EAD$, and $\angle ABC$ are shown in the design. What is the measure of $\angle ADE$?

   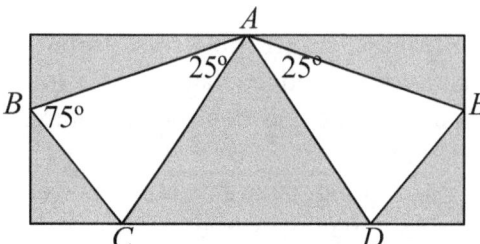

   A. 25°
   B. 50°
   C. 75°
   D. 80°
   E. 90°

167

Section 7.7
Proof and Logic

Geometry Theorems and Postulates

I have never seen the ACT use a real two column proof, but they can ask about some of the theorems. Do you remember proving that triangles are congruent using sides and angles, like "side-angle-side" (SAS)? **SSS, SAS, ASA,** and **AAS** prove triangles are congruent. That's any combination of three angles or sides except two:

      AAA only proves they're similar. (Think about it: not all 45-45-90 triangles are the same size.)
      SSA doesn't exist. (If it spells a bad word either frontwards or backwards, don't use it.)

Never forget the most important multiple-choice-geometry theorem:

"If it looks congruent, it probably is."

Example 1

In the figure to the right, $\angle ABC \cong \angle ADC$ and $\angle BAC \cong \angle DAC$.
Which of the following congruences must be true?

    A. $\overline{AB} \cong \overline{AC}$

    B. $\angle ABC \cong \angle ACD$

    C. $\overline{AD} \cong \overline{AC}$

    D. $\angle ACB \cong \angle ADC$

    E. $\overline{BC} \cong \overline{CD}$

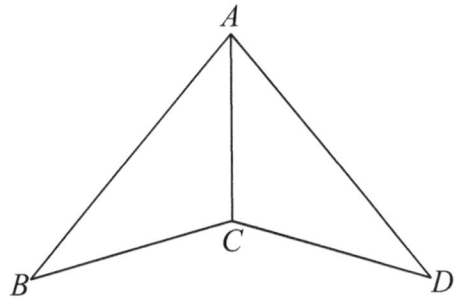

The two triangles must be congruent, or the question would be impossible. Therefore, all we need are pieces that match. $\overline{BC}$ and $\overline{CD}$ match, so they must be congruent. Choose **E**.

Conditionals, Contrapositives, and other big words

I'm sure you've noticed that the ACT loves to use big words to make easy stuff sound hard and confusing. Some questions don't require math at all, just common sense.
For example:

1. If it is raining, then the ground is wet.   ← This is a **conditional**.
    Assume this is true.

2. If the ground is wet, then it is raining.   ← This is a **converse**.
    False, maybe the sprinklers were on.

3. If it is NOT raining, then the ground is NOT wet.   ← This is an **inverse**.
    False, maybe the sprinklers were on.

4. If the ground is NOT wet, then it is NOT raining.   ← This is a **contrapositive**.
    True.

Statements 1 and 4 say the same thing, so they are both true. Statements 2 and 3 both say it backwards, so they are not *necessarily* true.

Example 2
Consider the following two logical statements.

  If the measure of ∠*A* is 30°, then the measure of ∠*B* is 45°.

  The measure of ∠*B* is **not** 45°.

If these statements are both true, then it follows that the measure of:
  A. ∠*A* is **not** 30°
  B. ∠*A* is 30°
  C. ∠*A* is 45°
  D. ∠*B* is 30°
  E. ∠*B* is **not** 30°

This is a logic problem, so we'll use some logic of our own. 30 goes with *A* and 45 goes with *B*, so eliminate answers **C**, **D**, and **E**. The real question is, "is ∠*A* 30°?" Since ∠*B* is **not** 45°, ∠*A* is **not** 30° either. Choose **A**.

7.7 Practice Problems: Answers on p. 185

What conclusion, if any, can be drawn from the following statements?
 1. If it is raining, the ground is wet.
  It is raining.
 2. All Squeebils have Trallz.
  Migulla does not have Trallz.
 3. If Mr. Rogers is in a bad mood, we have lots of homework.
  We have lots of homework.

 What postulate can prove the triangles congruent?
 4.
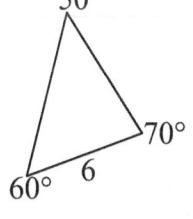 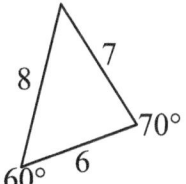
               5.
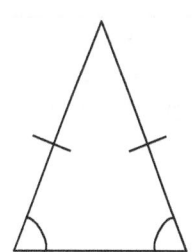 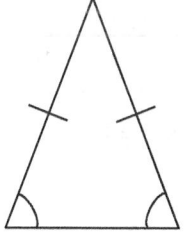

7.7 ACT Problems: Answers on p. 185

1. At Sunny Dayz Water Park, if a person is not more than 40 inches tall, then that person cannot ride the Typhoon water slide. If Daniel rode the Typhoon water slide at Sunny Dayz Water Park yesterday, then which of the following statements must logically be true?
 A. Daniel is at least 50 inches tall.
 B. Daniel is more than 40 inches tall.
 C. Daniel is exactly 40 inches tall.
 D. Daniel is less than 40 inches tall.
 E. Daniel is at most 30 inches tall.

2. Consider the following logical statements:
  All squares are parallelograms.
  ABCD is a square.
  WXYZ is a parallelogram.
If these statements are all true, which of the following must also be true?
 F. *ABCD* is a parallelogram.
 G. *WXYZ* is a square.
 H. *ABCD* is congruent to *WXYZ*.
 J. *ABCD* is NOT a parallelogram.
 K. *WXYZ* is NOT a square.

3. In the figure below, $\angle ACD \cong \angle BDC$, $\overline{AC}$ is perpendicular to $\overline{AD}$, and $\overline{BD}$ is perpendicular to $\overline{BC}$. Which of the following postulates can be used to prove that $\triangle ACD$ is congruent to $\triangle BDC$?

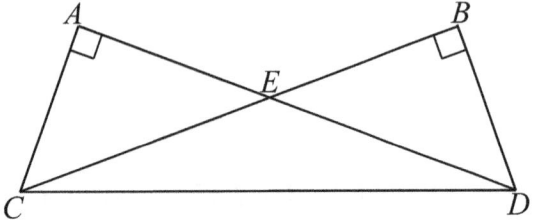

    A. SSS
    B. SAS
    C. AAA
    D. AAS
    E. SSA

Section 7.8
Circles

Vocabulary

Center – a point in the middle of a circle
Diameter – a line segment across the middle of a circle
Radius – a line segment from the center to the edge of a circle
Circumference – the length around a circle
Arc – a piece of the circumference

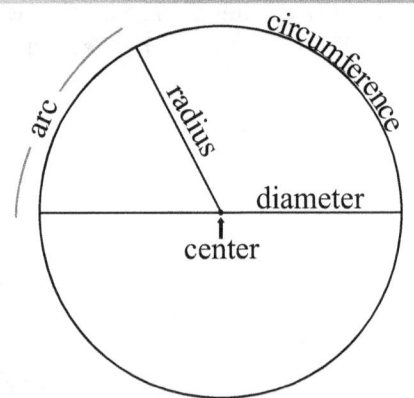

Tangent – a line that touches a circle in exactly one point
Secant – a line that touches a circle in two points, but does not pass through the center
Chord – a line segment with its endpoints on a circle, but that does not pass through the center

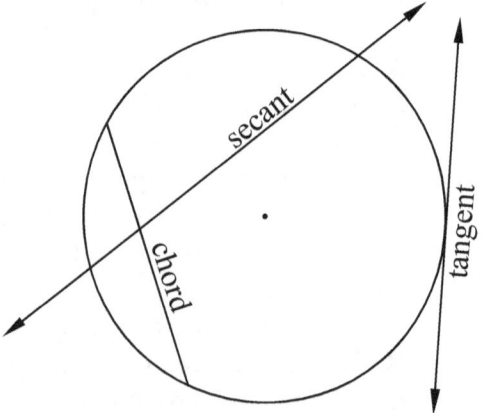

Angles in a Circle

Central angle – an angle with its vertex in the center of a circle
The measure of a central angle is equal to the measure of its arc.

Inscribed angle – an angle with its vertex on the edge of a circle
The measure of an inscribed angle is half the measure of its arc.

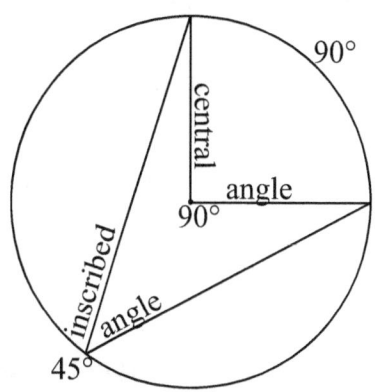

Example 1
In the figure below, the measure of $\overarc{AB}$ is 100°.
What is the measure of the inscribed angle, θ?

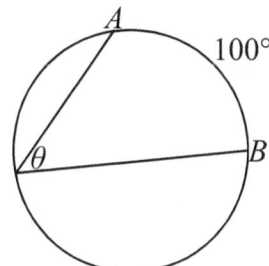

Since θ is an inscribed angle, its measure is half of the measure of the arc. $\theta = 100/2$
Also, θ is an acute angle; it must be smaller than 100°. $\theta = 50°$.

Arc Length and Area of a Sector

The same formula can find arc length and area of a sector: $\dfrac{part}{whole} = \dfrac{part}{whole}$

Arc length is part of the circumference.
Sector area is part of the area.

Example 2
In Figure 1 at right, find the length of $\overarc{AB}$.

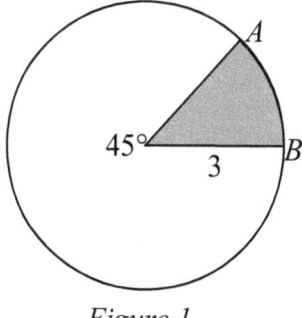

Figure 1

We start with our formula: $\dfrac{part}{whole} = \dfrac{part}{whole}$

We know that $\overarc{AB}$ is 45° and a whole circle is 360°.
Plug those numbers in. $\dfrac{45}{360} = \dfrac{part}{whole}$

$\overarc{AB}$ is part of the circle, and the length of a whole circle is the circumference.
$$C = 2\pi r = 6\pi$$

Plug in again. $\dfrac{45}{360} = \dfrac{x}{6\pi}$

Cross multiply. $270\pi = 360x$

Solve. $x = \dfrac{3\pi}{4}$

Example 3

In Figure 1 at right, find the area of the shaded region.

We start with our formula again:
$$\frac{part}{whole} = \frac{part}{whole}$$

We know that $\overset{\frown}{AB}$ is still 45° and a whole circle is still 360°.

Plug those numbers in.
$$\frac{45}{360} = \frac{part}{whole}$$

The shaded sector is part of the area of a whole circle.
$$A = \pi r^2 = 9\pi$$

Plug in again.
$$\frac{45}{360} = \frac{x}{9\pi}$$

Cross multiply.
$$405\pi = 360x$$

Solve.
$$x = \frac{9\pi}{8}$$

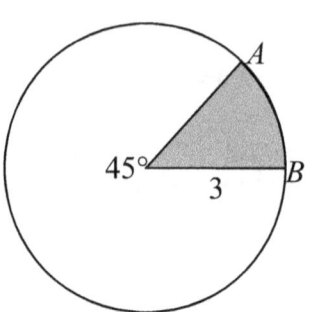

Figure 1

7.8 Practice Problems: Answers on p. 185-186

Find the length of $\overset{\frown}{ABC}$

1.

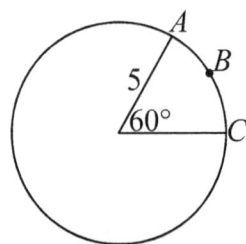

2.

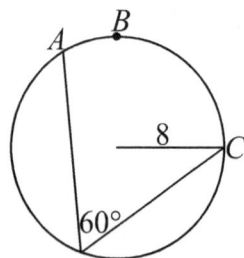

3.

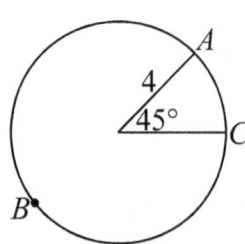

Find the area of the shaded region

4.

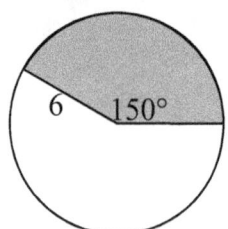

5.

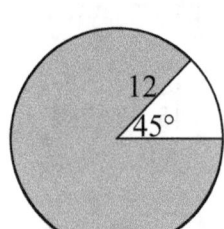

6.

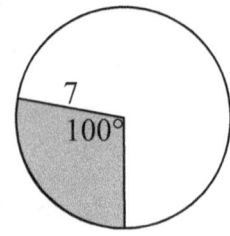

1. In the circle shown below, X is the center and lies on $\overline{AD}$ and $\overline{BC}$. Which of the following statements is NOT true?

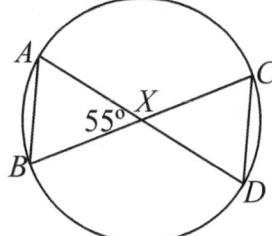

   A. $\overline{AX} \cong \overline{CX}$
   B. $\overline{AB} \cong \overline{CD}$
   C. $\overline{BX} \cong \overline{CD}$
   D. $\angle ABX$ measures 62.5°
   E. $\overparen{CD}$ measures 55°

2. What is the length, in feet, of a 200° arc of a circle whose circumference is 54 feet?
   F. 30 feet
   G. 34 feet
   H. 108 feet
   J. 160 feet
   K. 200 feet

3. If the radius of the circle below is 8 miles, and X is the center of the circle, approximately what is the length of arc $\overparen{YZ}$?

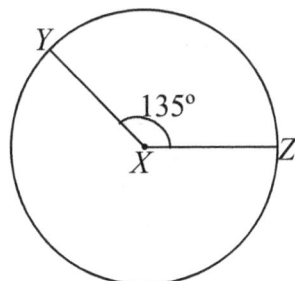

   A. 6 miles
   B. 16 miles
   C. 18.84 miles
   D. 42.67 miles
   E. 50.24 miles

4. If the radius of the circle below is 6 inches, what is the area of the shaded portion of the circle, in square inches?

   F. 6π
   G. 9π
   H. 12π
   J. 27π
   K. 36π

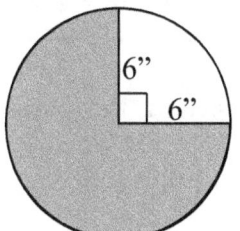

5. A plastic cover is being made for a feed bin. The bin has a diameter of 24 inches. The cover will open by a wedge shaped flap that forms a 60° angle at the center of the circular cover, as shown below. The flap will be secured by a zipper that runs along one side of the flap and around the arc. How long will the zipper need to be, to the nearest inch?

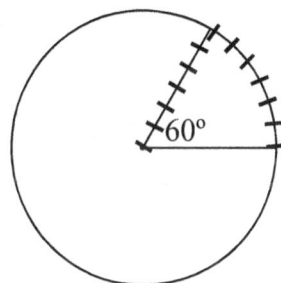

   A. 8 inches
   B. 12 inches
   C. 13 inches
   D. 20 inches
   E. 25 inches

Section 7.9
Area and Perimeter Formulas

Area is how much space is inside a shape. If you wanted to cover a 10 ft by 12 ft room with 1 foot square tiles, how many tiles would you need? You would need 10 rows of 12 tiles each, or 120 square foot tiles. That's area.

Perimeter is the distance around a shape. If you wanted to put a baseboard around a 10 ft by 12 ft room, how much baseboard would you need? 10 feet for the first wall, plus 12 ft for the second, plus 10 ft, plus 12 ft, which equals 44 feet of baseboard. That's perimeter.

Triangle Area	$A = \frac{1}{2}bh$	Circle Area	$A = \pi r^2$
Square Area	$A = s^2$	Circumference	$C = \pi d$ or $C = 2\pi r$
Rectangle Area	$A = lw$ or $A = bh$		
Parallelogram Area	$A = bh$	Parallelogram or Rectangle Perimeter	$P = 2(l + w)$
Trapezoid Area	$A = \frac{1}{2}h(b_1 + b_2)$ or $A = bh$ where b = the average base	Perimeter of Any Shape	Add up all the sides

Example 1
Find the area of a trapezoid with bases of 6 and 8 and a height of 3.

We can use the simple formula $A = bh$. If we use 8 as the base, we get the area of the big rectangle. If we use 6, we get the area of the small rectangle. The 8 is too big, and the 6 is too small. Find the average base, $\frac{8+6}{2} = 7$. Using 7 as the base gives a rectangle with the same area as the trapezoid, so $A = 7 \times 3 = 21$.

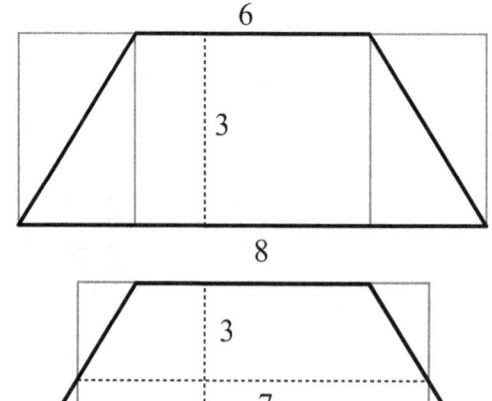

Example 2
What is the area of a circle that has a circumference of 24π cm?

We need the formulas for area and circumference of a circle. $\quad A = \pi r^2$ and $C = 2\pi r$.
They gave us circumference, so we plug it into the circumference formula. $\quad 24\pi = 2\pi r$
The only variable we can solve for is r. $\quad 12 = r$
They asked for area, not radius, so plug r into the area formula. $\quad A = \pi 12^2$
Solve. $\quad A = 144\pi$

Example 3

All the adjacent line segments in the figure at the right intersect in right angles. If each segment is 5 units long, what is the area, in square units, of the entire figure?

    A. 25
    B. 60
    C. 100
    D. 125
    E. 150

There is no formula for the area of a cross, so we need to cut the figure into shapes that we know. One way to cut it is to make five squares.

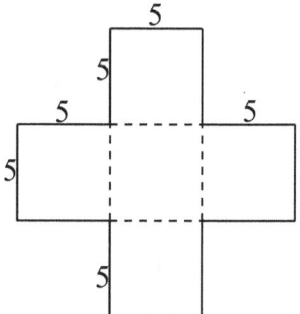

The area of one square is $A = s^2$ (or $A = lw$).
Plug in: $A = 5^2 = 25$

Since all five squares are congruent, they all have the same area, so we can multiply the area of one square times 5 to get the total area.

$A = 5 \times 25 = 125$
Choose **D**.

7.9 Practice Problems: Answers on p. 186

Find the area and perimeter of the following figures.

1.

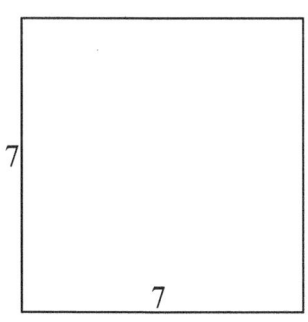

2.

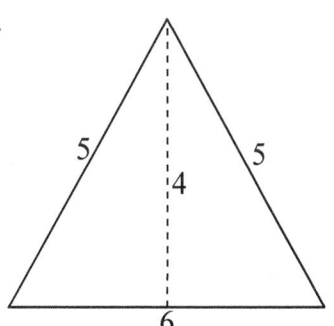

3.

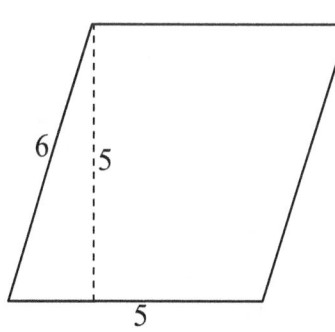

4.

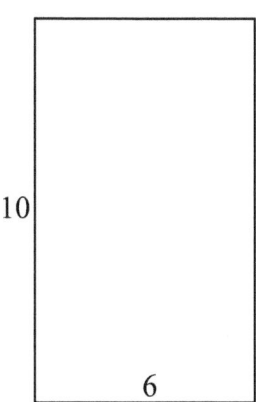

5.

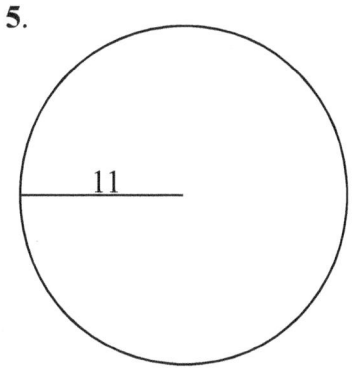

6.

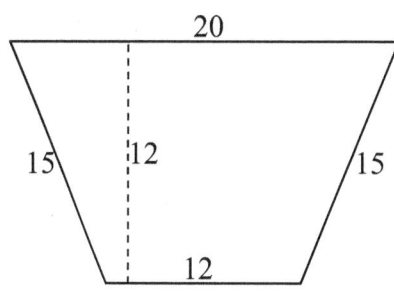

1. A pentagon has 1 side of length w m, 2 sides of length $(w + 2)$ m, 1 side of length $2w$ m, and 1 side of length 8 m. What is the perimeter, in meters, of the pentagon?
   A. $5w + 12$
   B. $5w + 8$
   C. $4w + 12$
   D. $4w + 8$
   E. $5w$

2. You have enough material to build a fence 48 meters long. If you use it all to enclose a square region, how many square meters will you enclose?
   F. 96
   G. 144
   H. 192
   J. 324
   K. 576

3. The following sketch shows a bedroom wall from the front. What is the area of the front surface of this wall, in square feet?

   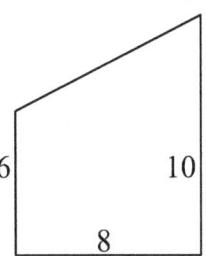

   A. 48
   B. 64
   C. 80
   D. 104
   E. 128

4. In the 8-sided figure below, adjacent sides meet at right angles and the lengths are given in centimeters. What is the perimeter of the figure, in centimeters?

   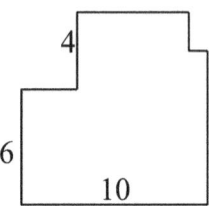

   F. 20
   G. 40
   H. 60
   J. 80
   K. 200

5. The figure below shows square $XYZO$ and also shows the circle centered at O. $\overline{XO}$ and $\overline{ZO}$ are radii of the circle. If the area of the square is 49 square units, what is the area of the circle, in square units?

   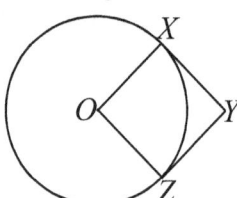

   F. 14π
   G. 21π
   H. 49π
   J. 98π
   K. 147π

Section 7.10
3-Dimensional Geometry

The two things we can find for a three-dimensional shape are **volume** and **surface area**.

Volume

If you have already memorized all the area formulas, you can use them to find volume. You only need to learn two more formulas to find the volume of anything except a sphere.

Prism, cube, box, cylinder, or any other shape that goes straight up and down.	$V = Ah$ A is the area of the base h is the height perpendicular to the base
Cone, pyramid, or any shape that goes up to a point	$V = \dfrac{Ah}{3}$ A is the area of the base h is the height perpendicular to the base
Sphere (Usually they will give you this formula if you need it)	$V = \dfrac{4}{3}\pi r^3$ Or you can estimate $V = \dfrac{1}{2}d^3$ (see Ex. 1)

Surface area

Surface area is just the areas of all of the surfaces added up. You don't need any new formulas.

Example 1

Find the volume of a sphere with a radius of 3 units.

If they give you the formula, just plug in.

$$V = \frac{4}{3}\pi r^3$$

$$V = \frac{4}{3}\pi (3)^3 = 36\pi$$

If they don't give the formula, estimate that the volume of a sphere is half the volume of a cube around the sphere.
If the radius is 3, the diameter is 6, and the sides of the cube are 6.

$$V \approx \frac{1}{2}d^3$$

$$V \approx \frac{1}{2}6^3 = 108$$

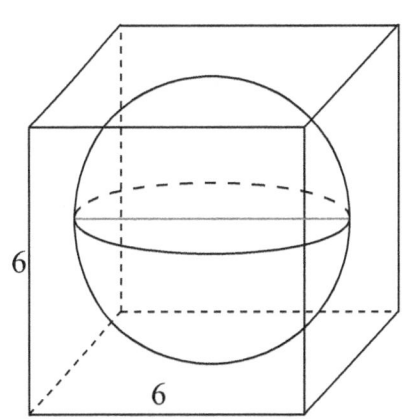

$36\pi \approx 113.04$ which is pretty close to 108. You should be able to guess the correct answer.

Example 2
Find the volume and the surface area of a cylinder with a diameter of 10 cm and a height of 10 cm.

We'll find the volume first. A cylinder goes straight up, so we need the formula $V = A h$.
The base is a circle, so $A = \pi r^2$.
If the diameter is 10, then the radius is 5.
Plug in the area and height to the volume formula.

$$A = \pi 5^2 = 25\pi$$
$$V = 25\pi(10)$$
$$V = 250\pi \text{ cm}^3$$

Next we'll find the surface area. If you were making a cylinder out of construction paper, how many pieces would you need, and what shape would they be? You would need two circles and a rectangle. Find the areas of the two circles and the rectangle.
We already found the circles.
The rectangle is length times width.
The width is 10, and the length is equal to the circumference of the circle.
Add up all three pieces.

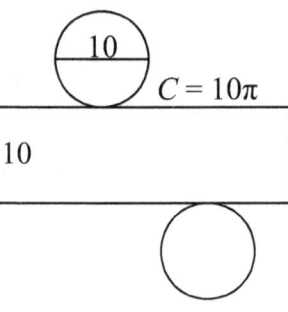

$$A = \pi 5^2 = 25\pi$$

$$C = 2\pi r = 10\pi$$
$$A = lw = 10\pi(10) = 100\pi$$
$$SA = 25\pi + 25\pi + 100\pi$$
$$SA = 150\pi \text{ cm}^2$$

7.10 Practice Problems: Answers on p. 187

1. What is the length of each side of a cube if the volume is 1331 in³?
2. What is the surface area of a rectangular prism with a length of 5 cm, width of 3 cm, and height of 2 cm?
3. What is the volume of a right circular cone if the radius is 4 m and the height is 9 m?
4. What is the height of a cylinder with a volume of 400π ft³ and a radius of 5 ft?
5. What is the surface area of a cylinder with a height of 8 in and a diameter of 6 in?
6. What is the volume of a sphere if the radius is 9 cm?
7. What is the volume of this box after it is assembled?

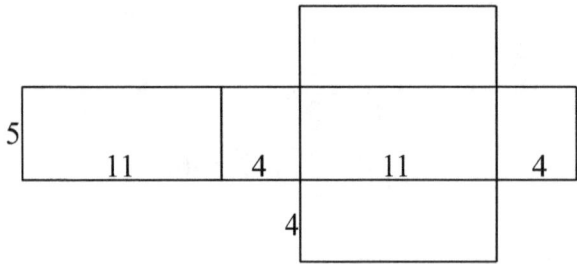

8. What is the volume of this box after it is assembled?

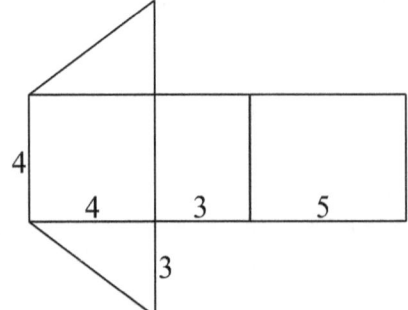

1. U Move It packing company needs a box with a volume of 172,800 in³. If the length is 60 in and the width is 48 in, what must be the height of the box?
   A. 24 in
   B. 36 in
   C. 48 in
   D. 60 in
   E. 72 in

2. A square pyramid has a surface area of 156 m². If the length of one side of the square base is 4 m, what is the area of each triangle in the pyramid?
   F. 35 m²
   G. 36 m²
   H. 38 m²
   J. 39 m²
   K. 40 m²

3. A swimming pool is 25 meters wide and 50 meters long. The deep end is 2 meters deep and the bottom is horizontal for 20 meters, until it slopes upward to a depth of 1.2 meters at the shallow end. What is the volume of water , in cubic meters, required to completely fill the pool?

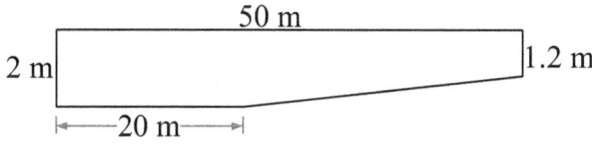

   A. 88 m³
   B. 440 m³
   C. 2000 m³
   D. 2200 m³
   E. 2500 m³

4. Find the surface area of the figure below.

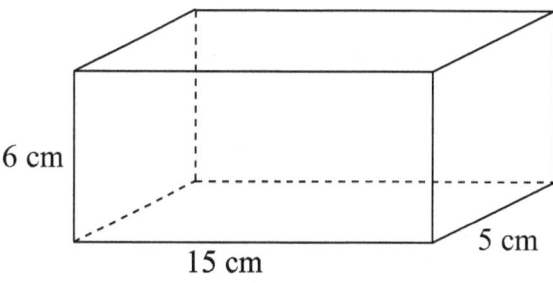

   A. 195 cm²
   B. 320 cm²
   C. 390 cm²
   D. 420 cm²
   E. 450 cm²

5. The volume, V, of a sphere is determined by the formula $V = \frac{4}{3}\pi r^3$, where r is the radius of the sphere. What is the volume, in cubic inches, of a sphere with a diameter 12 inches long?

   A. 144π
   B. 288π
   C. 460π
   D. 864π
   E. 2304π

Plane Geometry Review

1. Rhombus *ABCD* is shown below. If *BC* is 5 cm and *BD* is 8 cm, what is the length of $\overline{AC}$?

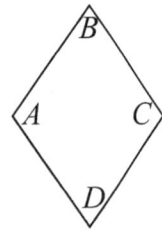

   A. 3 cm
   B. 4 cm
   C. 5 cm
   D. 6 cm
   E. 8 cm

2. In the figure below, *B* is on $\overline{AC}$ and $\overline{AC} \parallel \overline{DE}$. If $m\angle 3 = 80°$ and $m\angle 4 = 30°$, what is $m\angle 2$?

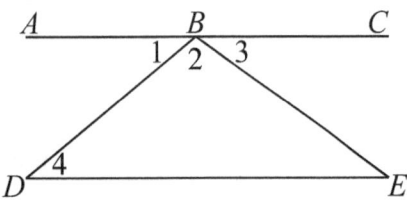

   F. 70°
   G. 60°
   H. 50°
   J. 30°
   K. Cannot be determined from the given information.

3. In the figure below, two transversals cross three parallel lines as shown. $\overline{AB}$ Is 5 inches, $\overline{DE}$ is 3 inches and, $\overline{EF}$ is 1 inch. What is the length of $\overline{BC}$?

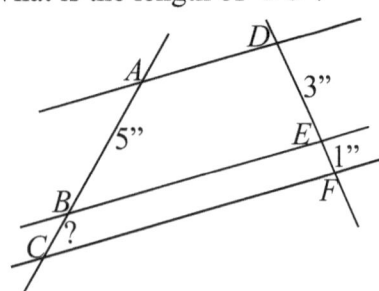

   A. 1"
   B. 1.5"
   C. 1.67"
   D. 2"
   E. 2.67"

4. Points *X*, *Y*, and *Z* Are on the circle below, whose center is at *O* and whose radius is 9 meters. The length of $\overline{BC}$ equals the length of $\overline{OC}$. What is the length, in meters, of minor arc *BC*?

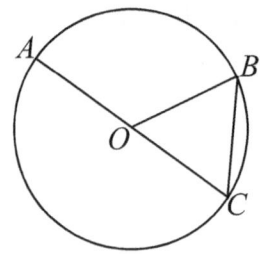

   F. 3π
   G. 6π
   H. 9π
   J. 12π
   K. 18π

5. In the isosceles right triangle shown below, *AB* = 8 feet. What is the length, in feet, of $\overline{AC}$?

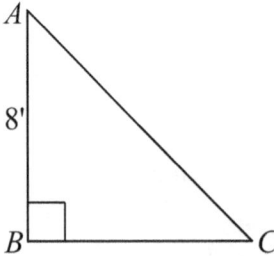

   A. 4
   B. 8
   C. 16
   D. $\sqrt{8}$
   E. $8\sqrt{2}$

6. In the figure below, with given side lengths, ∠V and ∠Z are right angles. What is the total area, in square units, of pentagon *VWXYZ*?

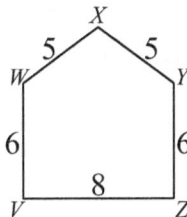

F. 48
G. 54
H. 58
J. 60
K. 72

7. In the figure below, all the points around rectangle *ADFI* are equally spaced around the perimeter. What is the ratio of the unshaded area to the shaded area?

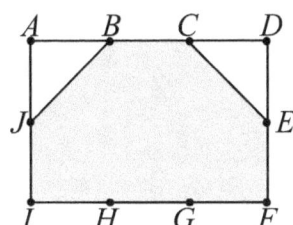

A. 1:5
B. 2:5
C. 1:6
D. 2:6
E. 5:6

8. The figure below is drawn on a grid with unit squares. Each vertex is at the intersection of 2 grid lines. What is the area of the hexagon, in square units?

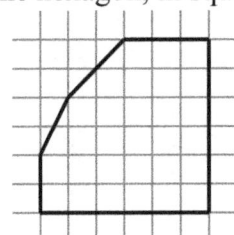

F. 29
G. 31
H. 33
J. 34
K. 36

9. In the regular hexagon below, $\overline{BD} \perp \overline{OC}$; *C* is the midpoint of $\overline{BD}$; and *O* is the midpoint of $\overline{AD}$. What is the degree measure of ∠*COD*?

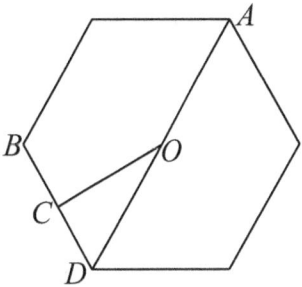

A. 15°
B. 30°
C. 45°
D. 60°
E. 150°

10. In △*ABC* below, point *D* is on $\overline{BC}$, $\overline{AB}$ is perpendicular to $\overline{BC}$, and $\overline{AD}$ bisects ∠*BAC*. If the measure of ∠*ACD* is 20°, what is the measure of ∠*ADC*?

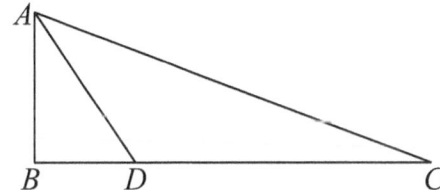

F. 35°
G. 55°
H. 70°
J. 125°
K. 160°

11. Consider the following two statements.

> If the area of *ABCD* is 16, then *AB* is 4.
> *AB* is NOT 4.

If these statements are both true, then it follows that the area of:
A. *ABCD* is 16
B. *ABCD* is NOT 16
C. *ABCD* is 4
D. *AB* is NOT 16
E. none of the above

181

12. There are 3 empty cubical containers with side lengths of 1, 2, and 4 feet respectively. How many fillings of the middle-sized container would be required to completely fill the largest container?

    F. 2

    G. 4

    H. 8

    J. 16

    K. 32

13. The ratio of the side lengths of a rectangle with an area of 78 m^2 is 3:1. Which of the following is closest to the length, in meters, of the longest side of the rectangle?

    A. 3

    B. 5

    C. 10

    D. 15

    E. 30

Answers on p. 187-188

14. The horizontal side of a right triangle is $(2x + 4)$ units long and the vertical side is $(3x - 6)$ units long. Which expression correctly represents the area, in square units, of the right triangle?

    F. $3x^2 - 6$

    G. $3x^2 + 6x - 12$

    H. $3x^2 - 12$

    J. $3x^2 + 12x - 12$

    K. $6x^2 + 12x - 12$

Answers and Explanations

Section 7.1 Basics of Plane Geometry
Practice Problems:
1. points $A, B, C, D, E, F, G,$ and H
2. $\overleftrightarrow{EF}$ and $\overrightarrow{HG}$
3. $\overline{AB}$, $\overline{CG}$, $\overline{DH}$, $\overline{AD}$, $\overline{EH}$, and $\overline{FG}$
4. $\overrightarrow{EA}$, $\overrightarrow{FB}$, $\overrightarrow{BC}$, and $\overrightarrow{DC}$
5. $\overleftrightarrow{EF}$ and $\overrightarrow{HG}$
6. $\overline{AB}$ & $\overline{AD}$, or $\overline{AD}$ & $\overline{DH}$, or $\overline{EH}$, & $\overline{DH}$, or $\overline{CG}$ & $\overline{FG}$
7. ABC or ABD or ABH or any three letters

7.1 ACT Problems:
1. **B.** $\overleftrightarrow{AB} \| \overleftrightarrow{CD}$ If two lines are both perpendicular to another line, they must be parallel to each other. On a sheet of notebook paper, any two blue lines are perpendicular to the red line and parallel to each other.
2. **F.** A line   A line is the only option that can cut all the way through a plane in both directions and leave even pieces. An angle doesn't leave equal pieces, and the others don't cut all the way through.
3. **D.** the length of segment $\overline{AB}$ If they wanted to name two points, they would have said "A and B," not "AB." Eliminate **A**. An angle can be named with one letter or three, not two. Eliminate **E**. You can't measure the length of a line or ray because they go on forever. Eliminate **B** and **C**. AB has to refer to the length of a segment.

Section 7.2 Angles
Practice Problems:
1. $\angle 1$ & $\angle 4$, $\angle 2$ & $\angle 3$, $\angle 5$ & $\angle 8$, and $\angle 6$ & $\angle 7$
2. $\angle 1$ & $\angle 5$, $\angle 2$ & $\angle 6$, $\angle 3$ & $\angle 7$, and $\angle 4$ & $\angle 8$
3. $\angle 3$ & $\angle 6$, $\angle 4$ & $\angle 5$
4. $\angle 1$ & $\angle 8$, $\angle 2$ & $\angle 7$
5. $\angle 3$ & $\angle 5$, $\angle 4$ & $\angle 6$
6. $\angle 1$ & $\angle 7$, $\angle 2$ & $\angle 8$
7. The pairs from #1–4 are congruent. #5 and 6 are supplementary.

7.2 ACT Problems:
1. **E.** 5600°   If m$\angle 3$ = 140°, then all of the obtuse angles are 140°. All of the acute angles are 180 − 140 = 40°. That means that m$\angle 2$ = 140° and m$\angle 5$ = 40°. $140 \times 40 = 5600$°.
2. **H.** 100°   Since both pairs of lines are parallel, all the acute angles are 80°. Therefore all of the obtuse angles are 100°. Angle z is obtuse, so it equals 100°.
3. **E.** $\angle 3 \cong \angle 4$ Angle 4 is the only angle on the bottom line, so we have to use it. $\overline{AB}$ is not part of $\angle 4$, so ignore it. $\angle 3$ is the only other acute angle made from $\overline{BC}$ and a parallel line like $\angle 4$, so they must be the congruent pair.
4. **H.** 8   First find the 4 angles in the lower right intersection. Then subtract from 180 to find the third angle in the middle triangle. Then find the 4 angles on the lower left. The angle in the middle top has to be 80 because its vertical angle is 80. The other four are 50 because all of the acute angles made with a parallel line and either transversal are 50. There are 8 50° angles.

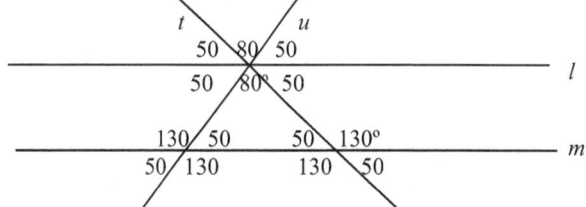

5. **A.** 220°   Since both pairs of lines are parallel, all the acute angles are 70°. Therefore all of the obtuse angles are 110°. Angles 1 and 2 are obtuse, so their sum is 110 + 110 = 220°.

Section 7.3 Quadrilaterals and Polygons
Practice Problems:
1. parallelogram
2. rectangle
3. $\angle B$ = 140°, $\angle C$ = 40°, $\angle D$ = 140°
4. 540°
5. 360°
6. trapezoid
7. 135°

7.3 ACT Problems:

1. **D.** All of the above A square and a rhombus have 4 equal sides, so their perimeters have to be 4 times a side. If two of the answers are right, we have to choose all of the above. The question asks what the figure *could* be. A trapezoid could have sides with lengths 2, 3, 3, and 4. The perimeter would be 12, which is 4 times 3, and 3 is the length of one side.

2. **F.** add up to 180° Any geometric figure must be named with the letters going around in order, like below. $\angle ABC$ and $\angle BCD$ have to come one after the other, so one is acute and the other is obtuse. They add up to 180°

3. **C.** *BDKI* The left and right sides are parallel, and only $\overline{BI}$ and $\overline{DK}$ can make the parallel top and bottom.

4. **K.** 500° The formula for the sum of the interior angles of a polygon is $180(n-2)$. For a pentagon, $n = 5$. $180(5-2)=540$ One angle is 40, so together the other four must be 500°.

5. **C.** 36° We know that the sum of the interior angles is 540° (see number 4 above), and since this is a regular pentagon, the 5 angles are all the same. $540 \div 5 = 108$ If the interior angles are 108°, then the exterior angles are $180 - 108 = 72$. Subtract from 180 to find the third angle in the triangle. $\angle X = 36°$

Section 7.4 Triangles
Practice Problems:

1. 60°
2. 25° & 130° or 77.5° & 77.5°
3. 96°
4. 30°
5. yes, because 5 + 11 > 13
6. no, because 2 + 9 is not greater than 11

7.4 ACT Problems:

1. **C.** 16°

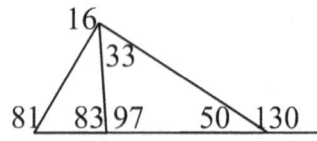

Supplementary angles add up to 180°, and the three angles in a triangle add up to 180°. Use those facts to find more angles until you can find the one you want.

2. **K.** I, II, and III Statement I is the easiest to see. The three angles in a triangle add up to 180. Angles 2 and 4 are congruent because they are vertical angles, so 2 can replace 4. 5 and 7 are also the same, and so are 10 and 12.

3. **B.** 2, 3, 6 The two short sides can't meet in the middle because 2 + 3 is only 5. The sum of the two short sides has to be longer than the long side.

4. **F.** The measure of $\angle X$ is always less than the measure of $\angle Y$. The biggest angle is always across from the biggest side; the smallest angle is always across from the smallest side; and the medium angle is always across from the medium side.

5. **E.** (10,5) If the 4 legs of the right triangles are all equal, then point E is twice as far to the right as it is up. (10,5) is the only set with the x coordinate twice as big as the y.

Section 7.5 Right Triangles
Practice Problems:

1. 17
2. 8
3. $\sqrt{2}$
4. $3\sqrt{2}$
5. $\sqrt{13}$
6. $2\sqrt{3}$
7. $8\sqrt{3}$

7.5 ACT Problems:

1. **E.** $\sqrt{193}$ Use the Pythagorean Theorem.
    $$7^2+12^2=c^2$$
    $$193=c^2$$
    $$\sqrt{193}=c$$

2. **J.** $6+6\sqrt{2}$ Since the triangle is isosceles, $\overline{YZ}=6$. Use the Pythagorean Theorem.
    $$6^2+6^2=c^2$$

$72 = c^2$

$6\sqrt{2} = c$

Or, if you remember 45–45–90 triangles, the legs are $x = 6$ and the hypotenuse is $x\sqrt{2} = 6\sqrt{2}$. Don't forget to add $\overline{YZ}$ and $\overline{XZ}$ to get $6 + 6\sqrt{2}$.

3. **D**. 60° 30–60–90 triangles have sides in the ratio shown. $\angle C$ is bigger than $\angle A$, so it's 60°, not 30°.

4. **G**. 17.9 feet Use the Pythagorean Theorem. $8^2 + 16^2 = c^2$

$320 = c^2$

$17.88 = c$

Round to 17.9 feet.

5. **C**. $5\sqrt{3}$ The diagram is a 45–45–90 triangle and a 30–60–90 triangle. Start with $\triangle ABC$. Either remember 45–45–90 triangles or use the Pythagorean Theorem. $x^2 + x^2 = (5\sqrt{2})^2$ The sides of $\triangle ABC$ are 5. Since the triangle on the right is 30–60–90,

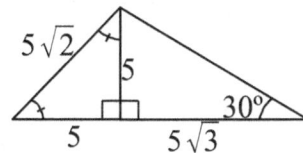

the hypotenuse is $2x$ and the longer leg is $x\sqrt{3} = 5\sqrt{3}$.

Section 7.6 Similar Triangles
Practice Problems:

1. $A = 58°$, $B = 46°$, $C = 76°$, $x = 10.5$, $Y = 46°$, $z = 12$
2. $B = 28°$, $C = 140°$, $X = 12°$, $y = 17.3$, $Y = 28°$, $z = 24$
3. $A = 77°$, $B = 43°$, $C = 60°$, $x = 101.8$, $z = 89.9$, $Z = 60°$
4. $A = 25°$, $Y = 96°$, $Z = 59°$, $b = 4.7$, $c = 4$
5. 10 inches

7.6 ACT Problems:

1. **D**. 11.3 in. Use ratios.

$\dfrac{9}{16} = \dfrac{x}{20}$ 9 lines up with 16 because

$180 = 16x$ they're both the shortest sides.

$11.25 = x$ 9 is across from x because they're both on the same triangle. Use the 20 because it's the other longest side.

2. **G**. 6 We can't find $\overline{CQ}$ directly because it is not a side of a triangle. Find $\overline{AC}$ instead.

Use ratios. $\dfrac{5}{15} = \dfrac{3}{x}$

$\overline{AC}$ is 9. Subtract $\overline{AC} - \overline{AQ} = \overline{CQ}$

$9 - 3 = 6$

3. **D**. 20 feet Use ratios.

$\dfrac{9}{15} = \dfrac{12}{x}$

$x = 20$

4. **G**. 100° The acute angles at A and D both have to equal 55°, and the acute angle at E has to equal 25°. Subtract the angles of either triangle from 180° to find $\angle ACE$. $180 - 55 - 25 = 100$.

5. **D**. 80° $\angle ADE$ is congruent to $\angle C$ on the other white triangle. Subtract to find $\angle C$. $180 - 75 - 25 = 80$. $\angle ADE$ must also be 80°.

Section 7.7 Proof and Logic
Practice Problems:

1. The ground is wet.
2. Migulla is not a Squeebil.
3. no conclusion (maybe a different teacher gave the homework)
4. ASA
5. AAS (remember that SSA doesn't exist)

7.7 ACT Problems:

1. **B**. Daniel is more than 40 inches tall. Any English teacher who reads this will cringe at the double negative. Not bigger than 40 can't ride means bigger than 40 can ride.

2. **F**. ABCD is a parallelogram. If all squares are parallelograms, then square $ABCD$ is a parallelogram.

3. **D**. AAS

$\angle ACD \cong \angle BCD$

$\angle CAD \cong \angle DBC$ because they're both 90°

$\overline{CD} \cong \overline{CD}$ because both triangles use the same line

We have two angles and one side, and answer **D** is the only one that uses two angles and a side.

Section 7.8 Circles
Practice Problems:

1. $\dfrac{5\pi}{3}$

2. $\dfrac{16\pi}{3}$

185

3. 7π

4. 15π

5. 126π

6. $\dfrac{245\pi}{18}$

7.8 ACT Problems:

1. C. $\overline{BX} \cong \overline{CD}$

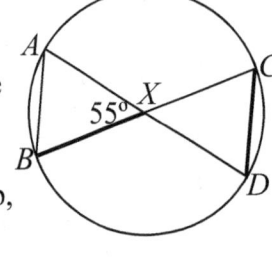

$\overline{BX}$ is touching the 55° angle, but $\overline{CD}$ is across from it. They don't match up, so they aren't congruent.

2. F. 30 feet

$\dfrac{part}{whole}=\dfrac{part}{whole}$

$\dfrac{200}{360}=\dfrac{x}{54}$

$x=\dfrac{200\times54}{360}=30$

3. C. 18.84 miles

$\dfrac{part}{whole}=\dfrac{part}{whole}$

$C=2\pi r=16\times3.14=50.24$

$\dfrac{135}{360}=\dfrac{x}{50.24}$

$x=\dfrac{135\times50.24}{360}=18.84$

4. J. 27π

$\dfrac{part}{whole}=\dfrac{part}{whole}$ $360-90=270°$

$A=\pi r^2=36\pi$

$\dfrac{270}{360}=\dfrac{x}{36\pi}$

$x=\dfrac{270\times36\pi}{360}=27\pi$

5. E. 25 inches

First, find the length of the arc.

$\dfrac{part}{whole}=\dfrac{part}{whole}$

$C=2\pi r=24\pi=75.36$

$\dfrac{60}{360}=\dfrac{x}{75.36}$

$x=\dfrac{60\times75.36}{360}=12.56$

Then add the length of the straight piece (the radius). $12.56+12=24.56\approx25$

Section 7.9 Area and Perimeter
Practice Problems:

1. area = 49, perimeter = 28

2. area = 12, perimeter = 16

3. area = 25, perimeter = 22

4. area = 60, perimeter = 32

5. area = 121π, perimeter = 22π

6. area = 192, perimeter = 62

7.9 ACT Problems:

1. A. $5w+12$ Add the five sides.

$w+(w+2)+(w+2)+2w+8=5w+12$

2. G. 144

A fence 48 meters long enclosing a square means the square has a perimeter of 48. Therefore, the sides of the square are 12 meters each. They asked for the area.

$A=s^2=12^2=144$

3. B. 64

The area of a trapezoid is $A=bh$. We have two heights, so we'll average them.

$h=\dfrac{6+10}{2}=8$

$A=bh=8\times8=64$

4. G. 40

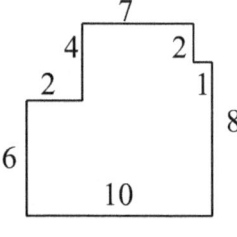

There are two ways to do this. First, you could make up numbers for the unlabeled sides. As long as the right sides add up to 10 and the top sides also add up to 10, the perimeter will be 40. Secondly, you could move the sides to make a rectangle. We only moved the sides, we didn't change any lengths, so the perimeter is the same. It's 40.

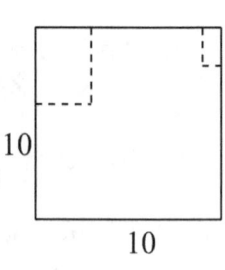

5. H. 49π

The area of a circle is $A=\pi r^2$ and the area of a square is $A=s^2$. Since the sides of the square are equal to the radius of the circle, $r^2=s^2$. $A=\pi r^2=49\pi$

186

Section 7.10 3-Dimensional Geometry
Practice Problems:
1. 11 in
2. 62 cm^2
3. 48π m^3
4. 16 ft
5. 66π in^2
6. 972π cm^3 = 3052.08 cm^3
7. 220
8. 24

7.10 ACT Problems:
1. **D**. 60 in $V = l \cdot w \cdot h$ Plug in:
   $172{,}800 = 60 \cdot 48 \cdot h$ Solve for h. $h = 60$
2. **F**. 35 m^2 A square pyramid is made up of
   one square base and four triangular sides.
   The square base has an area of $A = s^2 = 16$.
   The surface area is the square plus the four
   triangles. $SA = 16 + 4t$ Plug in $SA = 156$
   and solve for t. $156 = 16 + 4t$ $t = 35$
3. **D**. 2200 m^3 First, find the area of the
   cross-section shown. Cut it into a rectangle
   and a trapezoid.

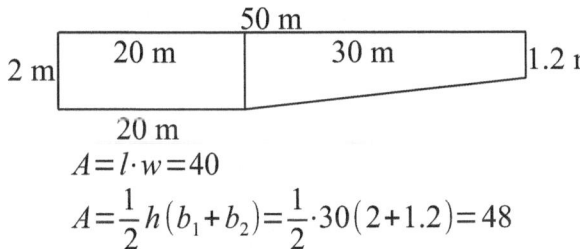

   $A = l \cdot w = 40$
   $A = \frac{1}{2} h (b_1 + b_2) = \frac{1}{2} \cdot 30 (2 + 1.2) = 48$

   Total area equals 88. They asked for
   volume, so we multiply by the dimension
   perpendicular to the cross-section, 25.
   $V = 88 \cdot 25 = 2200$
4. **C**. 390 cm^2 To find surface area, add up all
   the areas of the surfaces. The front and back
   are 90 each, the top and bottom are 75 each,
   and the left and right are 30 each. $SA = 90 + $
   $90 + 75 + 75 + 30 + 30 = 390$
5. **B**. 288π If the diameter is 12, the radius is
   6. $V = \frac{4}{3} \pi r^3 = \frac{4}{3} \pi \cdot 6^3 = 288 \pi$

Plane Geometry Review:
1. **D**. 6 cm Draw $\overline{BD}$ and $\overline{AC}$,
   making 4 right triangles. Use
   Pythagorean theorem to find
   the two horizontal pieces.

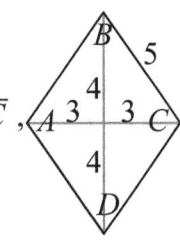

2. **F**. 70º
   $\angle 1 \cong \angle 4$
   because
   they are
   made
   from

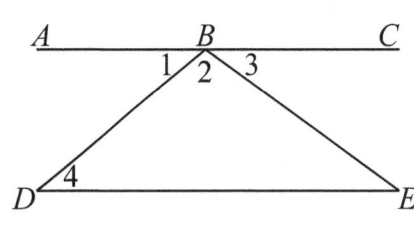

   parallel lines and the same transversal. $\angle 1$,
   $\angle 2$, and $\angle 3$ add up to 180º
3. **C**. 1.67" Set this up like similar triangles.
   $\frac{5}{?} = \frac{3}{1}$ Cross multiply. $? = \frac{5}{3} \approx 1.67$
4. **F**. 3π The radius is 9,
   and $\overline{BC}$ is equal to the
   radius. It's an
   equilateral triangle, so
   $\angle BOC = 60°$. To find
   arc length, use $C = 2\pi r$
   $= 18\pi$ and

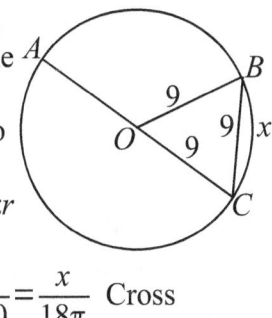

   $\frac{part}{whole} = \frac{part}{whole}$. $\frac{60}{360} = \frac{x}{18\pi}$ Cross
   multiply. $x = \frac{60 \times 18\pi}{360} = 3\pi$
5. **E**. 8$\sqrt{2}$ In an isosceles triangle, two sides
   are equal, so $BC = 8$ feet also. Use
   Pythagorean theorem: $8^2 + 8^2 = c^2 = 128$
   $c = \sqrt{128} = 8\sqrt{2}$
6. **J**. 60 First, break it into
   easy shapes. Make a
   rectangle and a triangle. Use
   Pythagorean theorem to find
   the height of the triangle.
   Find the two areas and add.
   $A_{rec} = l \cdot w = 8 \cdot 6 = 48$ and

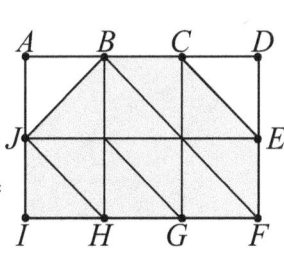

   $A_{tri} = \frac{1}{2} b \cdot h = \frac{1}{2} 8 \cdot 3 = 12$ $A = 48 + 12 = 60$
7. **A**. 1:5 Add lines so
   that all of the
   triangles are the
   same size. Count.
   Unshaded : shaded =
   2:10 = 1:5

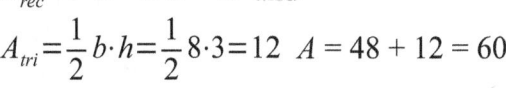

8. **G**. 31 Count the
   squares.

9. B. 30° Draw more lines. All of the central angles are congruent, and it takes 12 of them to go all the way around (360°).

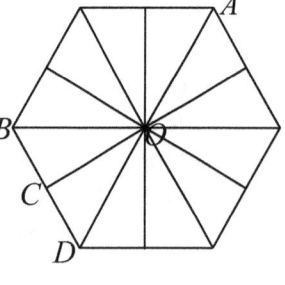

$$\angle COD = \frac{360}{12} = 30°$$

10. J. 125°

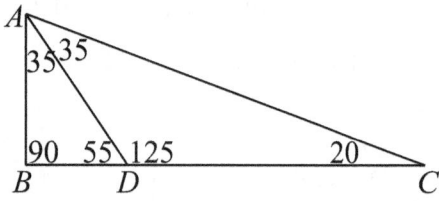

Label the angles that you know. Find $\angle BAC$ = 180 − 90 − 20 = 70. It's bisected, so the two halves are 35 each. $\angle BDA$ = 180 − 90 − 35 = 55. Then $\angle BDA$ = 180 − 55 = 125°

11. B. *ABCD* is NOT 16
If the area of *ABCD* is 16, then *AB* is 4.
AB is NOT 4.
The statements tell us that *AB* has to be 4 if *ABCD* has an area of 16. Since *AB* is NOT 4, then *ABCD*'s area can't be 16.

12. H. 8   Filling a container refers to the volume. Volume of a cube is $V = s^3$. The middle sized container has $V = 2^3 = 8$ and the large container has $V = 4^3 = 64$. How many 8's does it take to make 64? $8 \cdot x = 64$
$x = 8$.

13. D. 15
$A = l \cdot w$
$78 = 3x \cdot x = 3x^2$
$x = 5.099$
They asked for the long side, so $3x \approx 15$.

14. H. $3x^2 - 12$
$$A = \frac{1}{2} b \cdot h = \frac{1}{2}(2x+4)(3x-6)$$
$$A = (x+2)(3x-6)$$
$$A = 3x^2 - 6x + 6x - 12 = 3x^2 - 12$$

Chapter 8
Trigonometry

There are only four trig problems on the whole ACT. If you have never learned trig before, don't worry about it. Just guess on those 4. The good news about the trig problems is that they are usually NOT trick questions. If you know a little bit of trig, you should be able to work them easily.

Section 8.1
Triangle Ratios

Trigonometry is all about the ratios of the sides of right triangles. The three basic trig functions are also buttons on your calculator. You should memorize these three ratios.

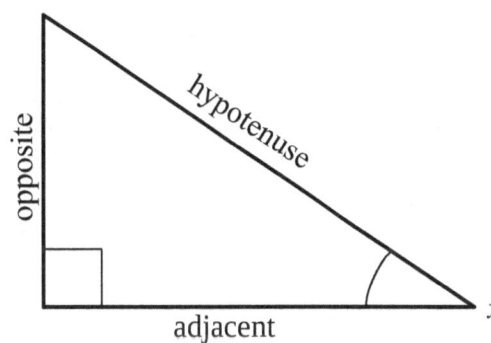

$$\sin x = \frac{\text{opposite}}{\text{hypotenuse}}$$

$$\cos x = \frac{\text{adjacent}}{\text{hypotenuse}}$$

$$\tan x = \frac{\text{opposite}}{\text{adjacent}}$$

Hypotenuse means the side across from the right angle; it's also the longest side. Opposite means across from the angle in the trig function. Adjacent means the side that's touching the angle in the trig function (other than the hypotenuse).

You can remember that **Sin** x = **O**pposite over **H**ypotenuse, **Cos** x = **A**djacent over **H**ypotenuse, and **Tan** x = **O**pposite over **A**djacent my memorizing **SOH CAH TOA**, or "**S**ome **O**ld **H**ippie **C**aught **A**nother **H**ippie **T**rippin' **O**n **A**cid."

Example 1
Find the sine, cosine, and tangent of ∠A.

We need to find the hypotenuse.
Use the Pythagorean Theorem.
$3^2 + 4^2 = c^2$ so $c = 5$.
The side opposite ∠A is 3.
The side adjacent to ∠A is 4.
The hypotenuse is 5.

$$\sin A = \frac{\text{opp}}{\text{hyp}} = \frac{3}{5} \qquad \cos A = \frac{\text{adj}}{\text{hyp}} = \frac{4}{5} \qquad \tan A = \frac{\text{opp}}{\text{adj}} = \frac{3}{4}$$

Sec, csc, and cot

The other 3 less common trig functions are secant, cosecant, and cotangent. They are just the first three trig functions flipped upside-down.

$$\sec x = \frac{1}{\cos x} = \frac{\text{hyp}}{\text{adj}} \qquad \csc x = \frac{1}{\sin x} = \frac{\text{hyp}}{\text{opp}} \qquad \cot x = \frac{1}{\tan x} = \frac{\text{adj}}{\text{opp}}$$

Cotangent is easy to remember; it's just tangent upside-down. Secant and cosecant are sine and cosine upside-down, but the "co-'s" are backwards. You would think that <u>co</u>secant would go with <u>co</u>sine, but it doesn't. <u>Co</u>secant is sine upside-down, and secant is <u>co</u>sine upside-down.

Example 2

Find the 6 trig functions of ∠X.

The side opposite ∠X is 2.
The side adjacent to ∠X is 1.
The hypotenuse is $\sqrt{5}$.

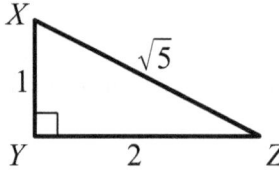

$$\sin X = \frac{\text{opp}}{\text{hyp}} = \frac{2}{\sqrt{5}} \qquad \cos X = \frac{\text{adj}}{\text{hyp}} = \frac{1}{\sqrt{5}} \qquad \tan X = \frac{\text{opp}}{\text{adj}} = 2$$

$$\csc X = \frac{\text{hyp}}{\text{opp}} = \frac{\sqrt{5}}{2} \qquad \sec X = \frac{\text{hyp}}{\text{adj}} = \sqrt{5} \qquad \cot X = \frac{\text{adj}}{\text{opp}} = \frac{1}{2}$$

Radians and degrees

Radians and degrees are two ways to measure an angle, just like feet and meters are two ways to measure a length. You can use either unit, but don't mix them up because they're not the same. If you are using the trig buttons on your calculator, make sure your calculator is set to the correct unit.

A whole circle in degrees is 360°, and a whole circle in radians is 2π. $\qquad \dfrac{\text{degrees}}{360} = \dfrac{\text{radians}}{2\pi}$

You can use ratios to convert from one unit to the other.

Special right triangles

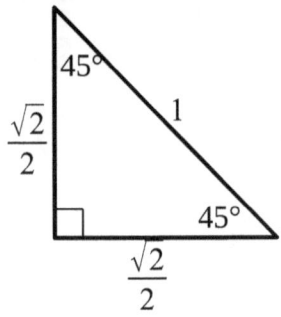

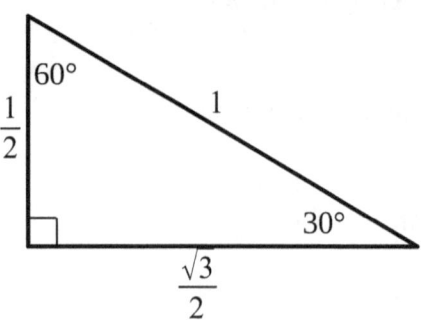

$$\sin 45 = \frac{\sqrt{2}}{2} \qquad \cos 45 = \frac{\sqrt{2}}{2} \qquad \tan 45 = 1 \qquad\qquad \sin 30 = \frac{1}{2} \qquad \cos 30 = \frac{\sqrt{3}}{2} \qquad \tan 30 = \frac{1}{\sqrt{3}}$$

$$\sin 60 = \frac{\sqrt{3}}{2} \qquad \cos 60 = \frac{1}{2} \qquad \tan 60 = \sqrt{3}$$

The unit circle

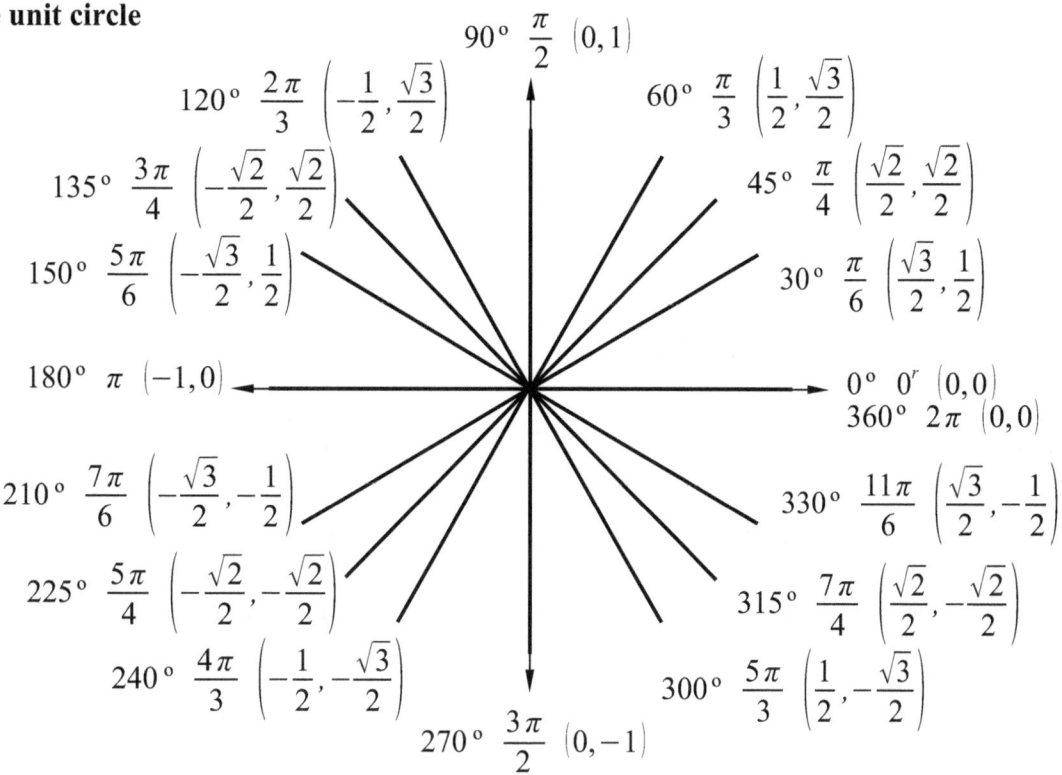

The unit circle has a radius of 1, so the triangle drawn for any angle on it has a hypotenuse of 1. The opposite side is equal to y, and the adjacent side is equal to x. For any angle, θ, on the unit circle with coordinates (x, y) The trig ratios can be found from those coordinates.

$$\sin\theta = y \qquad\qquad \cos\theta = x \qquad\qquad \tan\theta = \frac{y}{x}$$

That's great if you like to memorize. If you're not so good at memorization, you can use a calculator.

Example 3
$\cos 30° = ?$

If you have a unit circle drawn or memorized,
you know that the answer is $\frac{\sqrt{3}}{2}$.

The rest of us need a calculator.

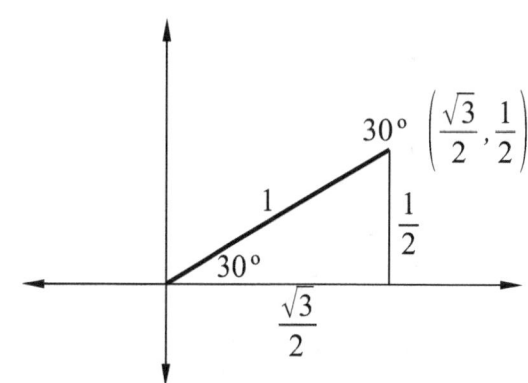

First, make sure your calculator is in degrees.
Type $\cos(30)$ ENTER.
You should get approximately 0.866025404.
Square both sides and convert to fraction.

$$(\cos(30))^2 = \frac{3}{4}$$

Square root both sides by hand.

$$\cos(30) = \frac{\sqrt{3}}{\sqrt{4}} = \frac{\sqrt{3}}{2}$$

Example 4
Solve for x.

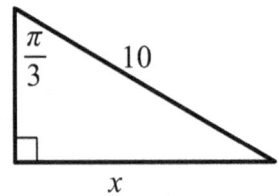

The two sides that are labeled are the opposite, x, and the hypotenuse, 10.

That means we need sine.

To solve for x, multiply both sides by 10.

Find $\sin\left(\dfrac{\pi}{3}\right)$ on a calculator or the unit circle.

(Make sure your calculator is in radians.)
Simplify.

$$\sin\left(\frac{\pi}{3}\right)=\frac{x}{10}$$

$$10\cdot\sin\left(\frac{\pi}{3}\right)=x$$

$$10\left(\frac{\sqrt{3}}{2}\right)=x$$

$$5\sqrt{3}=x$$

8.1 Practice Problems: Answers on p. 204

Find the 6 trig ratios for each angle.
1. $30°$

2. $45°$

3. $\dfrac{\pi}{2}$

4. $150°$

5. $\dfrac{5\pi}{4}$

6. $\angle A$

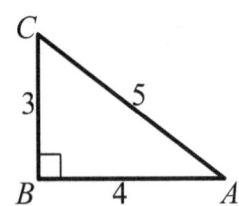

7. $\angle X$

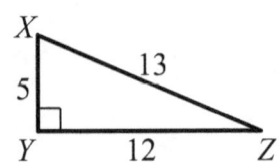

Solve for x.
8.

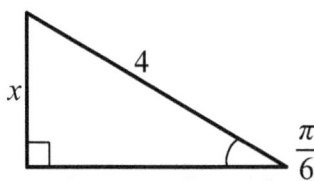

9.

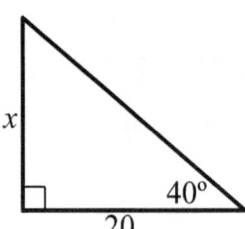

10.

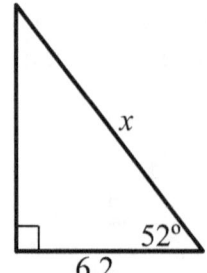

1. The side lengths of a right-angled triangle are 51 meters, 140 meters, and 149 meters, as shown below. What is the tangent of the smallest of the triangles 3 angles?

   A. $\dfrac{51}{140}$

   B. $\dfrac{51}{149}$

   C. $\dfrac{140}{51}$

   D. $\dfrac{140}{149}$

   E. $\dfrac{149}{51}$

   51 m

   149 m

   140 m

2. For the right triangle below, which of the following expressions is equal to $\cos \theta$?

   F. $\dfrac{x}{y}$

   G. $\dfrac{y}{x}$

   H. $\dfrac{x}{z}$

   J. $\dfrac{y}{z}$

   K. $\dfrac{z}{x}$

   x z θ y

3. In right triangle ABC illustrated below, the sine of $\angle A$ is $\dfrac{2}{3}$. What is the cosine of $\angle A$?

   A. $\dfrac{2}{3}$

   B. $\dfrac{3}{2}$

   C. $\sqrt{5}$

   D. $\dfrac{2}{\sqrt{5}}$

   E. $\dfrac{\sqrt{5}}{3}$

   C B A

4. In the figure below, $\tan \angle BAC = ?$

   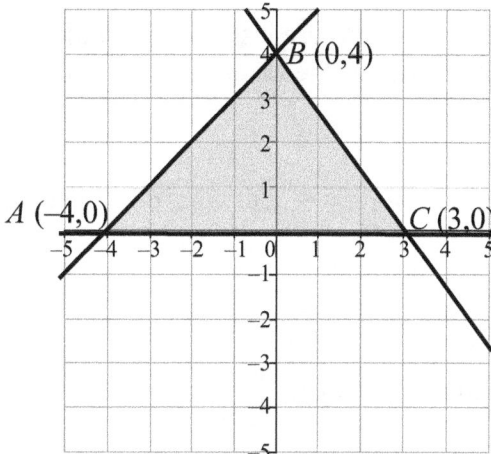

   F. 1

   G. $\dfrac{5}{7}$

   H. $\dfrac{5}{4\sqrt{2}}$

   J. $\dfrac{\sqrt{2}}{2}$

   K. $\dfrac{\sqrt{2}}{7}$

5. Linlee wants to find the distance directly across a lake to a tree on the opposite shore. She turns 90° from the tree and walks in a straight line for 150 yards. She turns to face the tree and measures the angle from her path to her line of sight to be 25°. How far, to the nearest yard, is the tree from Linlee's starting point?

   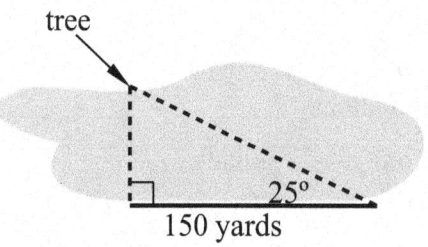

   tree

   25°

   150 yards

   A. 63 yards
   B. 70 yards
   C. 75 yards
   D. 136 yards
   E. 322 yards

Section 8.2
Inverse Trigonometric Functions

Inverse trig functions are just a way to undo trig functions. If you know the angle, sin, cos, and tan help you find the sides of the triangle. If you know the sides but want to find the angle, you need inverse trig functions.

We know that $\sin A = \frac{3}{5}$, but what does A equal?

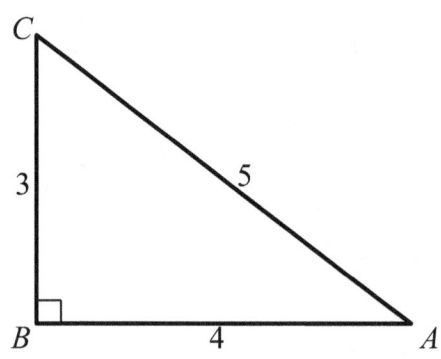

$$\sin A = \frac{3}{5}$$

$$\sin^{-1}(\sin A) = \sin^{-1}\left(\frac{3}{5}\right)$$

$$A = \sin^{-1}\left(\frac{3}{5}\right)$$

On a calculator, $\sin^{-1}\left(\frac{3}{5}\right) = 36.869898$ degrees or 0.643501 radians.

$$\sin^{-1}\left(\frac{\text{opp}}{\text{hyp}}\right) = \theta \qquad \cos^{-1}\left(\frac{\text{adj}}{\text{hyp}}\right) = \theta \qquad \tan^{-1}\left(\frac{\text{opp}}{\text{adj}}\right) = \theta$$

$$\sin^{-1}\left(\frac{3}{5}\right) = 36.8 \qquad \cos^{-1}\left(\frac{4}{5}\right) = 36.8 \qquad \tan^{-1}\left(\frac{3}{4}\right) = 36.8$$

Example 1
In the triangle pictured, $\theta = ?$

The labeled sides are the opposite and the hypotenuse, so we'll use sine.

$$\sin\theta = \frac{8}{16}$$

$$\sin^{-1}\left(\frac{8}{16}\right) = \theta$$

Use a calculator.

$$\theta = 30° \text{ or } \theta = \frac{\pi}{6} \text{ radians.}$$

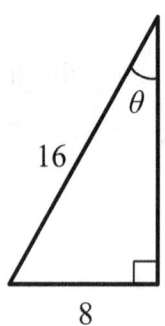

194

Example 2

If $180° < \theta < 270°$ and $\cos\theta = -0.2$, $\theta = ?$

Use inverse cosine to solve for θ.

$\cos\theta = -0.2$

$\cos^{-1}(-0.2) = \theta$

Make sure your calculator is in degrees.

$\cos^{-1}(-0.2) = 101.536959$

We found an answer, but it's not between 180° and 270°.

To move around the unit circle, add or subtract 180° or 360°.

$360 - 101.5 = 258.5$

258.5° is between 180° and 270°.

$\theta = 258.5°$

Check your answer.

$\cos(258.5) = -0.2$

8.2 Practice Problems: Answers on p. 204

1. $\angle X = ?$

2. $\angle Z = ?$

3. $\angle A = ?$

4. $\angle C = ?$

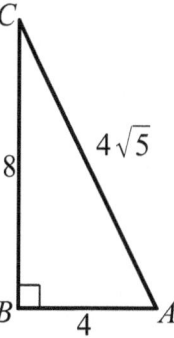

5. If $0° < \theta < 180°$ and $\cos\theta = 0.4$, $\theta = ?$

6. If $90° < \theta < 270°$ and $\sin\theta = -\dfrac{1}{2}$, $\theta = ?$

8.2 ACT Problems: Answers on p. 204

1. For the triangle below, $a = 6$ and $h = 11$. Find θ.

   A. 30°
   B. 33°
   C. 48°
   D. 57°
   E. 90°

2. The label on a ladder warns that for every four feet up a wall the ladder reaches, the base must be 1 foot from the wall. What is the measure of the angle between the ladder and the ground if this rule is followed?

   F. 14°
   G. 44°
   H. 62°
   J. 70°
   K. 76°

3. A plane flying at an altitude of 30,000 feet approaches the airport. If the plane is 600,000 feet away from the airport, what is the angle of depression from the plane to the airport?

   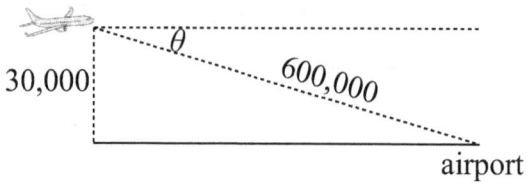

   A. 2.9°
   B. 16.7°
   C. 26.6°
   D. 28.7°
   E. 30°

Section 8.3
Trigonometric Identities and Laws

Since the six trig functions are ratios of the same three numbers, when they are multiplied, divided, or otherwise combined, numbers can cancel and leave you with other trig functions.

For example, $\dfrac{\sin\theta}{\cos\theta}=\dfrac{\frac{\text{opp}}{\text{hyp}}}{\frac{\text{adj}}{\text{hyp}}}=\dfrac{\text{opp}}{\text{hyp}}\cdot\dfrac{\text{hyp}}{\text{adj}}=\dfrac{\text{opp}}{\text{adj}}=\tan\theta$

For the ACT, you don't really need to know all of these identities, you just need to know that they exist.

Quotient Identities	Double-Angle Identities	Half-Angle Identities
$\tan x=\dfrac{\sin x}{\cos x}$	$\sin(2x)=2\sin x\cos x$	$\sin\left(\dfrac{x}{2}\right)=\pm\sqrt{\dfrac{1-\cos x}{2}}$
$\cot x=\dfrac{\cos x}{\sin x}$	$\cos(2x)=\cos^2 x-\sin^2 x$	$\cos\left(\dfrac{x}{2}\right)=\pm\sqrt{\dfrac{1+\cos x}{2}}$
Pythagorean Identities	$\cos(2x)=1-2\sin^2 x$	$\tan\left(\dfrac{x}{2}\right)=\pm\sqrt{\dfrac{1-\cos x}{1+\cos x}}$
$\sin^2 x+\cos^2 x=1$	$\cos(2x)=2\cos^2 x-1$	$\tan\left(\dfrac{x}{2}\right)=\dfrac{1-\cos x}{\sin x}$
$\tan^2 x+1=\sec^2 x$	$\tan(2x)=\dfrac{2\tan x}{1-\tan^2 x}$	$\tan\left(\dfrac{x}{2}\right)=\dfrac{\sin x}{1+\cos x}$
$1+\cot^2 x=\csc^2 x$		

Angle-Sum and -Difference Identities	Sum Identities	
$\sin(x+y)=\sin x\cos y+\cos x\sin y$	$\sin x+\sin y=2\sin\left(\dfrac{x+y}{2}\right)\cos\left(\dfrac{x-y}{2}\right)$	
$\sin(x-y)=\sin x\cos y-\cos x\sin y$	$\sin x-\sin y=2\cos\left(\dfrac{x+y}{2}\right)\sin\left(\dfrac{x-y}{2}\right)$	
$\cos(x+y)=\cos x\cos y-\sin x\sin y$	$\cos x+\cos y=2\cos\left(\dfrac{x+y}{2}\right)\cos\left(\dfrac{x-y}{2}\right)$	
$\cos(x-y)=\cos x\cos y+\sin x\sin y$	$\cos x-\cos y=-2\sin\left(\dfrac{x+y}{2}\right)\sin\left(\dfrac{x-y}{2}\right)$	
$\tan(x+y)=\dfrac{\tan x+\tan y}{1-\tan x\tan y}$		
$\tan(x-y)=\dfrac{\tan x-\tan y}{1+\tan x\tan y}$		

Example 1

$$\frac{\sin(2x)}{\tan x} = ?$$

    A. $\sin^2 x$

    B. $\cos^2 x$

    C. $2\sin^2 x$

    D. $2\cos^2 x$

    E. $\tan^2 x$

method 1 - identities

Use a double-angle identity

and a quotient identity.

Divide and simplify.

$$\sin(2x) = 2\sin x \cos x$$

$$\tan x = \frac{\sin x}{\cos x}$$

$$\frac{\sin(2x)}{\tan x} = \frac{2\sin x \cos x}{\dfrac{\sin x}{\cos x}}$$

$$\frac{\sin(2x)}{\tan x} = 2\sin x \cos x \cdot \frac{\cos x}{\sin x}$$

$$\frac{\sin(2x)}{\tan x} = 2\cos^2 x$$

The answer is **D**.

method 2 - calculator

Make up a number for x.

Don't use anything on the unit circle.

Plug in.

$$x = 20$$

$$\frac{\sin(2 \cdot 20)}{\tan 20} = 1.766044443$$

Plug 20 into the answer choices.

    A. $\sin^2 20 = 0.116977778$

    B. $\cos^2 20 = 0.883022222$

    C. $2\sin^2 20 = 0.233955557$

    D. $2\cos^2 20 = 1.766044443$

    E. $\tan^2 20 = 0.132474331$

The answer is **D**.

My calculator is in degrees. If yours is in radians, you will get different numbers, but answer **D** will still match.

Note: To type $\sin^2 20$ in a calculator, type "sin(20)^2" for most calculators. If that doesn't work for yours, type "(sin 20)^2" instead.

Law of sines and law of cosines

The other two formulas ACT can ask you to use are the law of sines and law of cosines. These two formulas are NOT limited to right triangles. They work on ANY triangle. However, every time I have seen ACT ask one of these problems, they have given the formulas.

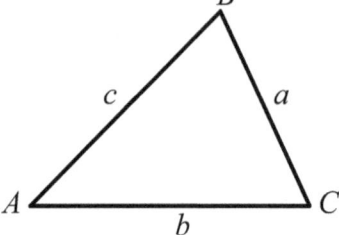

These formulas apply for any triangle with sides of length a, b, and c that are <u>opposite</u> angles A, B, and C, respectively.

law of sines
$$\frac{\sin A}{a}=\frac{\sin B}{b}=\frac{\sin C}{c}$$

law of cosines
$$c^2=a^2+b^2-2ab\cos C$$

The law of sines is easier. You should always try to use it first. If you don't have the right variables to make the law of sines work, then the law of cosines will be set up perfectly.

Example 2
For the triangle shown to the right, find a.

Try the law of sines.
$$\frac{\sin 45}{a}=\frac{\sin B}{6}=\frac{\sin C}{5}$$
No matter which two out of three ratios we choose, we still have 2 variables, so we can't solve it.

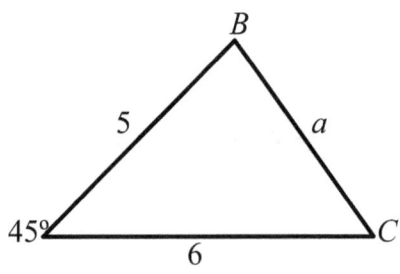

Try the law of cosines. (Switch the C and c with A and a.)
$$a^2=5^2+6^2-2(5)(6)\cos 45$$
$$a^2=18.57$$
$$a=4.3$$

8.3 Practice Problems: Answers on p. 204

1. $\cos^2 a-5+\sin^2 a=?$

2. $\cos(2\theta)+\sin^2\theta=?$

3. $(1-\sin^2 x)(\tan^2 x+1)=?$

4. $\csc\theta\cdot\sin^2\theta\cdot\sec\theta=?$

In the figure below,
5. $A=?$

6. $B=?$

7. $b=?$

1. Whenever $\dfrac{2\cos x \sin x}{\cos^2 x + 1 - \sin^2 x}$ is defined, it simplifies to:

    A. $\sin x$

    B. $\cos x$

    C. $\tan x$

    D. $\cos x \sin x$

    E. $\dfrac{2}{\cos x - \sin x}$

2. In $\triangle ABC$, shown below, the measure of $\angle A$ is 33°, the measure of $\angle C$ is 45°, and $\overline{AC}$ is 11 units long. Which of the following expressions represents the length, in units, of $\overline{AB}$?

    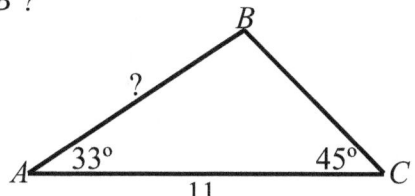

    F. $\dfrac{11 \sin 33°}{\sin 45°}$

    G. $\dfrac{11 \sin 45°}{\sin 33°}$

    H. $\dfrac{11 \sin 33°}{\sin 102°}$

    J. $\dfrac{11 \sin 45°}{\sin 102°}$

    K. $\dfrac{11 \sin 102°}{\sin 45°}$

3. In the figure below, the two triangles share a common side. Find $\cos(\alpha + \beta)$.
    (Note: $\cos(x+y) = \cos x \cos y - \sin x \sin y$ for all x and y.)

    A. $\dfrac{132}{625}$

    B. $\dfrac{44}{125}$

    C. $\dfrac{3}{5}$

    D. $\dfrac{99}{100}$

    E. $\dfrac{5}{4}$

4. The sides of an acute triangle measure 20 in, 22 in, and 24 in, respectively. Which of the following equations, when solved for θ, gives the measure of the largest angle of the triangle?

    F. $\dfrac{\sin \theta}{20} = \dfrac{1}{22}$

    G. $\dfrac{\sin \theta}{22} = \dfrac{1}{24}$

    H. $\dfrac{\sin \theta}{24} = \dfrac{1}{20}$

    J. $20^2 = 22^2 + 24^2 - 2(22)(24)\cos\theta$

    K. $24^2 = 20^2 + 22^2 - 2(20)(22)\cos\theta$

Section 8.4
Trigonometry Graphs

If you don't have a graphing calculator, it can be useful to know the general shapes of the trig functions, especially the three basic ones. If you do have a graphing calculator, make sure you know how to graph on it!

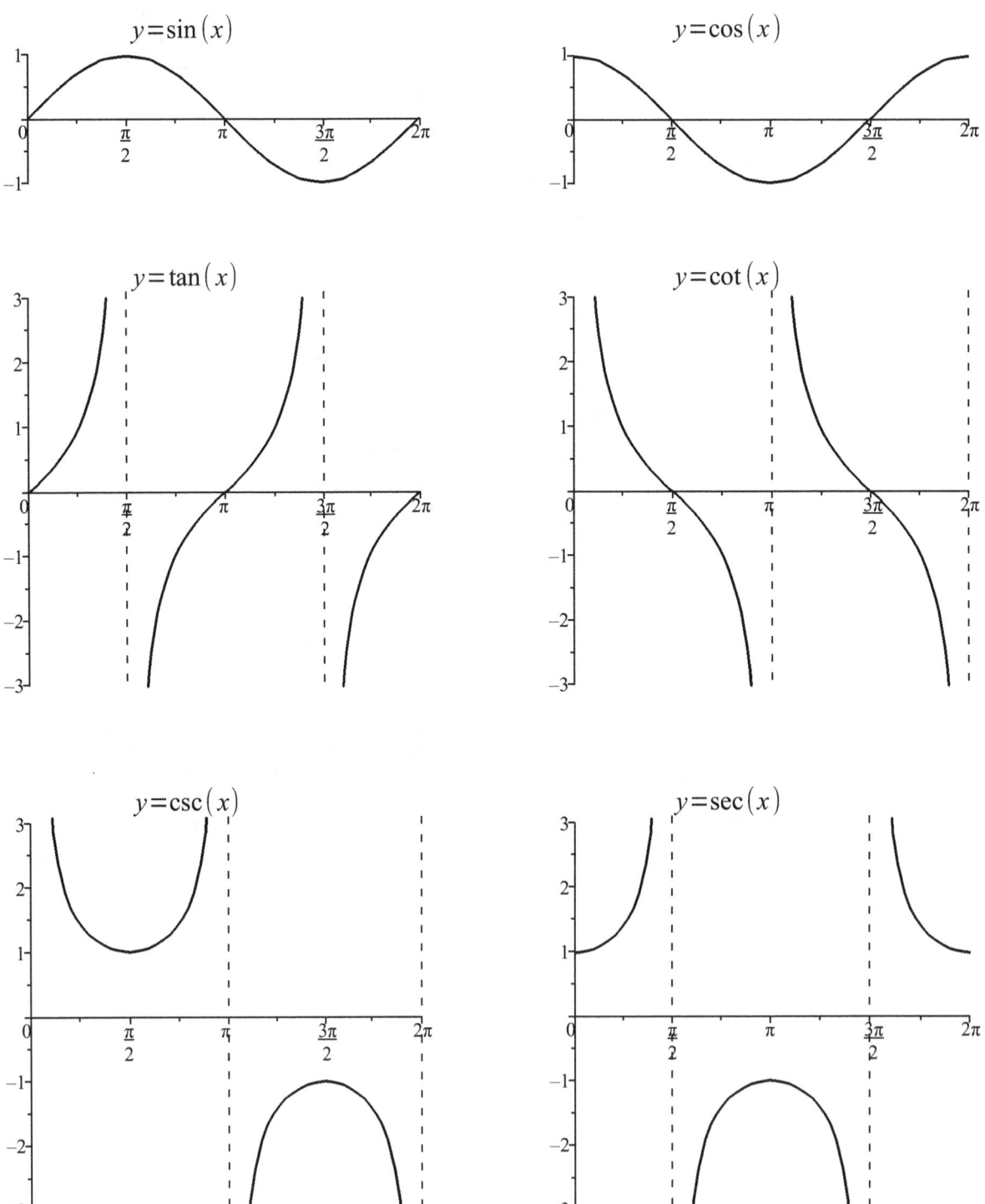

Amplitude and period

The **amplitude** is how high up the function goes from it's centerline. It is the number multiplied in front of a function, such as the "3" in $y=3\sin(x)$. Technically, only sine and cosine have amplitude because the range of the other trig functions goes up to infinity. However, that number in front of the function can tell you the "steepness" of tan or cot, and it can tell you the vertices of sec and csc.

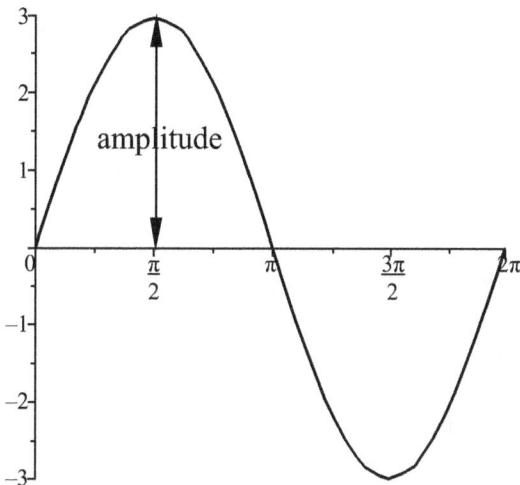

Amplitude is always positive, but if the number in front of the function is negative, the graph will be flipped vertically.

The **period** of a function is how far it goes before it repeats. Imagine you could copy part of a function and then paste, paste, paste. Would you get the whole graph again? Normally, the periods of sin, cos, sec, and csc are 2π. The periods of tan and cot are π.

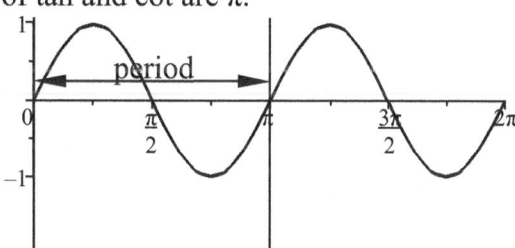

The figure above is $y=\sin(2x)$. To find the period, set the part of the function multiplied by x equal to the normal period of the particular trig function. $(2x)=2\pi$. Solve to find the period: $x=\pi$.

Example 1

Find the amplitude and period of $y=\dfrac{2}{3}\sin\left(\dfrac{2x+\pi}{3}\right)$.

The amplitude is the number in front of sine or cosine. $\text{amplitude}=\dfrac{2}{3}$

To find the period, set the variable and its coefficients equal to the normal period. Ignore the stuff added. $\left(\dfrac{2x}{3}\right)=2\pi$

Solve. $\text{period}=3\pi$

8.4 Practice Problems: Answers on p. 205

Find the amplitude and period.

1. $y=2\sin(2x)$

2. $y=4\cos\left(\dfrac{x}{2}\right)$

3. $y=-\dfrac{1}{3}\sin(2\pi x)$

4. $y=2\tan(x)$

5. $y=-\cos\left(\dfrac{x+\pi}{2}\right)$

6. $y=4\tan(3x)$

8.4 ACT Problems: Answers on p. 205

1. Compared to the graph of $y=\cos x$, the graph of $y=2\cos x$ has:
 A. half the amplitude and the same period.
 B. half the period and the same amplitude.
 C. double the amplitude and the same period.
 D. double the period and the same amplitude.
 E. double the amplitude and double the period.

2. The graph below shows two periods of which of the following functions?

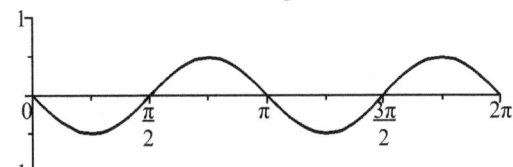

 F. $y=-\dfrac{1}{2}\sin(2x)$

 G. $y=2\cos\left(\dfrac{x}{2}\right)$

 H. $y=-2\sin(2x)$

 J. $y=-\dfrac{1}{2}\cos(2x)$

 K. $y=\dfrac{1}{2}\sin\left(\dfrac{x}{2}\right)$

3. The graphs of $y_1(x)=a_1\sin(b_1 x)$ and $y_2(x)=a_2\cos(b_2 x)$ are shown below. If a_1, b_1, a_2, and b_2 are all positive real numbers, which of the following statements is true?

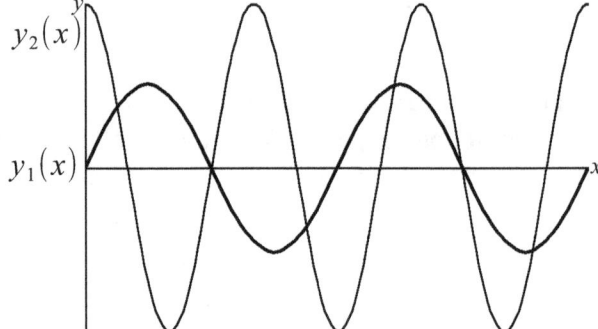

 A. $a_1<a_2$ and $b_1<b_2$
 B. $a_1<a_2$ and $b_1>b_2$
 C. $a_1>a_2$ and $b_1<b_2$
 D. $a_1>a_2$ and $b_1>b_2$
 E. $a_1<a_2$ and $b_1=b_2$

4. What is the product of the amplitude and period of the function $y=-3\cos(2\theta)$?
 F. -6π
 G. -3π
 H. $\dfrac{3}{2}\pi$
 J. 3π
 K. 6π

Trigonometry Review

1. In the figure below, what is the sine of ∠BAC?

   A. $\dfrac{2}{3}$

   B. $\dfrac{2}{\sqrt{13}}$

   C. $\dfrac{3}{2}$

   D. $\dfrac{3}{\sqrt{13}}$

   E. $\dfrac{3}{5}$

   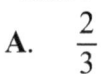

2. When measured from a point on the ground that is 45 feet from the base of a flagpole, the angle of elevation to the top of the flagpole is 48°, as shown below. What is the height, in feet, of the flagpole?

   F. $45\sin 48°$
   G. $45\cos 48°$
   H. $45\tan 48°$
   J. $45\sec 48°$
   K. $45\cot 48°$

   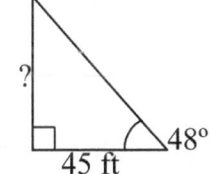

3. A land owner recorded the measurements around a pond as shown in the figure below. If he wants to calculate the length of the pond, which of the following formulas would be most applicable?

   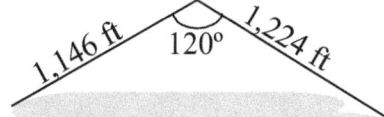

   A. $a^2+b^2=c^2$

   B. $A=\dfrac{1}{2}bh$

   C. $\sin(\theta)=\dfrac{\text{opp}}{\text{hyp}}$ and $\cos(\theta)=\dfrac{\text{adj}}{\text{hyp}}$

   D. $\dfrac{\sin A}{a}=\dfrac{\sin B}{b}=\dfrac{\sin C}{c}$

   E. $c^2=a^2+b^2-2ab\cos C$

Answers on p. 205

4. A 16-foot ladder is leaning against the wall of a house. If the base of the ladder is placed 6 feet from the house, and the wall makes a 90° angle with the ground, which of the following expressions gives the angle between the ladder and the ground?

   F. $\sin^{-1}\left(\dfrac{6}{16}\right)$

   G. $\cos^{-1}\left(\dfrac{6}{16}\right)$

   H. $\tan^{-1}\left(\dfrac{6}{16}\right)$

   J. $\sec^{-1}\left(\dfrac{6}{16}\right)$

   K. $\csc^{-1}\left(\dfrac{6}{16}\right)$

5. The graph below shows one period of which of the following functions?

   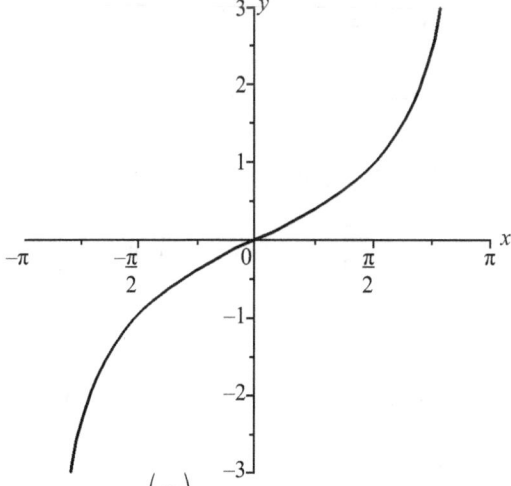

   A. $y=\tan\left(\dfrac{x}{2}\right)$

   B. $y=\tan(x)$

   C. $y=\tan(2x)$

   D. $y=\tan(3x)$

   E. $y=\tan(2\pi x)$

Answers and Explanations

Section 8.1 Triangle Ratios
Practice Problems:

1. $\sin 30° = \dfrac{1}{2}$, $\cos 30° = \dfrac{\sqrt{3}}{2}$,

   $\tan 30° = \dfrac{1}{\sqrt{3}}$ or $\dfrac{\sqrt{3}}{3}$, $\csc 30° = 2$,

   $\sec 30° = \dfrac{2}{\sqrt{3}}$, $\cot 30° = \sqrt{3}$

2. $\sin 45° = \dfrac{\sqrt{2}}{2}$, $\cos 45° = \dfrac{\sqrt{2}}{2}$, $\tan 45° = 1$,

   $\csc 45° = \sqrt{2}$, $\sec 45° = \sqrt{2}$, $\cot 45° = 1$

3. $\sin \dfrac{\pi}{2} = 1$, $\cos \dfrac{\pi}{2} = 0$, $\tan \dfrac{\pi}{2} = $ undefined,

   $\csc \dfrac{\pi}{2} = 1$, $\sec \dfrac{\pi}{2} = $ undefined, $\cot \dfrac{\pi}{2} = 0$

4. $\sin 150° = \dfrac{1}{2}$, $\cos 150° = -\dfrac{\sqrt{3}}{2}$,

   $\tan 150° = -\dfrac{1}{\sqrt{3}}$, $\csc 150° = 2$,

   $\sec 150° = -\dfrac{2}{\sqrt{3}}$, $\cot 150° = -\sqrt{3}$

5. $\sin \dfrac{5\pi}{4} = -\dfrac{\sqrt{2}}{2}$, $\cos \dfrac{5\pi}{4} = -\dfrac{\sqrt{2}}{2}$,

   $\tan \dfrac{5\pi}{4} = 1$, $\csc \dfrac{5\pi}{4} = -\sqrt{2}$,

   $\sec \dfrac{5\pi}{4} = -\sqrt{2}$, $\cot \dfrac{5\pi}{4} = 1$

6. $\sin A = \dfrac{3}{5}$, $\cos A = \dfrac{4}{5}$, $\tan A = \dfrac{3}{4}$,

   $\csc A = \dfrac{5}{3}$, $\sec A = \dfrac{5}{4}$, $\cot A = \dfrac{4}{3}$

7. $\sin X = \dfrac{12}{13}$, $\cos X = \dfrac{5}{13}$, $\tan X = \dfrac{12}{5}$,

   $\csc X = \dfrac{13}{12}$, $\sec X = \dfrac{13}{5}$, $\cot X = \dfrac{5}{12}$

8. $x = 2$
9. $x = 16.78$
10. $x = 10.07$

8.1 ACT Problems

1. **A.** $\dfrac{51}{140}$ $\quad$ tangent $= \dfrac{\text{opposite}}{\text{adjacent}} = \dfrac{51}{140}$

2. **J.** $\dfrac{y}{z}$ $\quad$ cosine $= \dfrac{\text{adjacent}}{\text{hypotenuse}} = \dfrac{y}{z}$

---

3. **E.** $\dfrac{\sqrt{5}}{3}$ Find the missing side using the Pythagorean theorem.

   cosine $= \dfrac{\text{adjacent}}{\text{hypotenuse}} = \dfrac{\sqrt{5}}{3}$

4. **F.** 1 Trig ratios are for right triangles. Ignore line $\overline{BC}$ and use the x- and y-axes. Opposite = 4 and adjacent = 4, so $\tan A = 1$.

5. **B.** 70 $\quad \tan 25 = \dfrac{x}{150}$ Cross-multiply.

   $150 \cdot \tan 25 = x \approx 69.946$

Section 8.2 Inverse Trigonometric Functions
Practice Problems

1. $\angle X = 30°$
2. $\angle Z = 60°$
3. $\angle A = 63.4°$
4. $\angle C = 26.6°$
5. $\theta = 66.4°$
6. $\theta = 210°$

8.2 ACT Problems

1. **D.** 57° $\quad \cos\theta = \dfrac{6}{11}$ $\quad \theta = \cos^{-1}\left(\dfrac{6}{11}\right) = 57$

2. **K.** 76° $\quad \tan\theta = \dfrac{4}{1}$ $\quad \theta = \tan^{-1}(4) = 76$

3. **A.** 2.9° $\quad \sin\theta = \dfrac{30{,}000}{600{,}000}$

   $\theta = \sin^{-1}\left(\dfrac{30{,}000}{600{,}000}\right) = 2.9$

Section 8.3
Practice Problems

1. -4
2. $\cos^2\theta$
3. 1
4. $\tan\theta$
5. $A = 45.5°$
6. $B = 67.5°$
7. $b = 6.2$

8.3 ACT Problems

1. **C.** $\tan x$ $\quad \dfrac{2\cos x \sin x}{\cos^2 x + 1 - \sin^2 x}$

204

$\sin^2 x + \cos^2 x = 1$ so $\cos^2 x = 1 - \sin^2 x$.
Substitute.

$$\frac{2\cos x \sin x}{\cos^2 x + (1 - \sin^2 x)} = \frac{2\cos x \sin x}{\cos^2 x + \cos^2 x}$$

Simplify. $\dfrac{2\cos x \sin x}{2\cos^2 x}$ Reduce. $\dfrac{\sin x}{\cos x}$ Use

a quotient identity. $\dfrac{\sin x}{\cos x} = \tan x$

Or plug in a number and match decimals.

2. **J.** $\dfrac{11 \sin 45°}{\sin 102°}$ First find the missing angle.

$B = 180 - 33 - 45 = 102$ Now use the law of

sines. $\dfrac{\sin 102°}{11} = \dfrac{\sin 45°}{x}$ Cross-multiply.

3. **B.** $\dfrac{44}{125}$ Use the given formula.

$\cos(x+y) = \cos x \cos y - \sin x \sin y$ Plug in.

$\cos(\alpha + \beta) = \dfrac{24}{25} \cdot \dfrac{15}{25} - \dfrac{7}{25} \cdot \dfrac{20}{25} = \dfrac{44}{125}$

4. **K.** $24^2 = 20^2 + 22^2 - 2(20)(22)\cos\theta$ Since
we only have sides and no angles, we have
to use the law of cosines. The law of sines
needs at least one angle given. The problem
asked for the biggest angle, which must be
across from the biggest side. That means θ
is across from 24, so 24 must be at the front
of the equation.

Section 8.4 Trigonometry Graphs
Practice Problems
1. amplitude: 2 period: π
2. amplitude: 4 period: 4π
3. amplitude: $\dfrac{1}{3}$ period: 1
4. amplitude: none period: π
5. amplitude: 1 period: 4π (ignore "$+\pi$")
6. amplitude: none period: $\dfrac{\pi}{3}$

8.4 ACT Problems
1. **C.** double the amplitude and the same
period The 2 in front of the function only
affects the amplitude, not the period.

2. **F.** $y = -\dfrac{1}{2}\sin(2x)$ The negative sign in

the front flips the graph down, the $\dfrac{1}{2}$

shrinks the amplitude so the graph only

goes up to $y = \dfrac{1}{2}$, and the $(2x)$ changes

the period to π by $(2x) = 2\pi \rightarrow x = \pi$.

3. **A.** $a_1 < a_2$ and $b_1 < b_2$ y_1 is shorter than
y_2, therefore $a_1 < a_2$. The period of y_1 is
longer than the period of y_2, therefore
$\dfrac{2\pi}{b_1} > \dfrac{2\pi}{b_2}$ and $b_1 < b_2$.

4. **J.** 3π The amplitude of $y = -3\cos(2\theta)$
is 3. The period is $(2\theta) = 2\pi \rightarrow \theta = \pi$.
The product is 3π.

Trigonometry Review
1. **B.** $\dfrac{2}{\sqrt{13}}$ $\sin\theta = \dfrac{\text{opp}}{\text{hyp}}$ Opposite is 2. Find
hypotenuse using the Pythagorean
Theorem. $c = \sqrt{2^2 + 3^2} = \sqrt{13}$.

2. **H.** $45\tan 48°$ The two labeled sides are
the opposite and adjacent, so we must be
using tan. $\tan 48° = \dfrac{?}{45}$

3. **E.** $c^2 = a^2 + b^2 - 2ab\cos C$ The triangle is
not a right triangle, so we can't use the
Pythagorean Theorem. We are given 2 sides
and an angle and asked for the third side.
Only the law of cosines uses three sides and
one angle.

4. **G.** $\cos^{-1}\left(\dfrac{6}{16}\right)$ The ground is the adjacent

side and the ladder is the hypotenuse.

$\cos\theta = \dfrac{\text{adj}}{\text{hyp}} = \dfrac{6}{16}$

5. **A.** $y = \tan\left(\dfrac{x}{2}\right)$ Normally, the period of tan

is π, but the period of this graph is 2π. Set
the part of the equation multiplied by x
equal to the normal period and solve.

$\dfrac{x}{2} = \pi \rightarrow x = 2\pi$

Chapter 9
Calculator Use

One of the most common questions both parents and students ask me is, "What calculator should I use on the ACT?" The answer is almost always, "A familiar one."

Of course, your calculator must be legal for use on the ACT. Check ACT's website for the most up-to-date information.

In general, if your calculator can solve equations and give answers in terms of variables, it's probably not legal. For example, if you tell the calculator that $A = lw$, and the calculator can tell you that $l = A/w$, that calculator is too smart to be allowed on the ACT.

However, if your calculator only tells you numbers, it's probably fine. For example, if you tell the calculator that $A = lw$, $l = 5$, $w = 2$, and it tells you that A is 10, that's probably an acceptable calculator.

In addition to being legal for the test, your calculator should, in my opinion, be able to understand order of operations, fractions, exponents, roots, and trigonometry functions, at minimum. A simple test is to type 1+2×3 in the calculator. If it says the answer is 7, the calculator understands order of operations. If it says the answer is 9, get a new calculator.

In this chapter, I explain some of the most useful functions of three common calculators. These are by no means the only acceptable calculators. You are welcome to use a different one, but I still want you to know how to use its features. The most important thing is to **use a calculator**. The ACT is timed. If your calculator can work a problem faster than you can, *use it!*

TI-30X

This calculator is often required by schools, so many of my students already own it. If you already have one, feel free to keep using it. It's not the best calculator, but it is a good calculator.

Features:

- One of the most useful but overlooked features is the delete/insert button. When you mistype something, do not clear it. Use the arrow keys to move on top of the mistake, then retype, use [DEL] to delete a character, or use INS, [2nd] [DEL], to insert characters.
- The ANS, [2nd][(-)], key, near the ENTER button, can save you from remembering and retyping long numbers.
    - 500 – ANS will subtract the previous answer from 500.
- This calculator is excellent with fractions and mixed numbers. The [A b/c] button separates the whole number (if needed), numerator, and denominator.

    - $\frac{4}{5}$ is typed 4 [A b/c] 5

    - $3\frac{1}{2}$ is typed 3 [A b/c] 1 [A b/c] 2

- F◄►D, [2nd][PRB], converts between fraction and decimal.
- Ab/c◄►d/e, [2nd][A b/c] converts between mixed number and improper fraction.
- Use the [^] button for exponents.
    - 2^3 is typed 2 [^] 3
    - 5^2 is typed 5 [^] 2 or 5 [x^2]
- Use $^x\sqrt{}$, [2nd][^] for roots.
    - $\sqrt[3]{8}$ is typed 3 [2nd][^] 8
    - $\sqrt{16}$ is typed 2 [2nd][^] 16 or [2nd][x^2] 16 [ )]
- When the calculator opens a parentheses for something like a square root, close it to end the square root.
    - $\sqrt{9}+1$ is typed [2nd][x^2] 9 [ )] + 1

- Use SCI/ENG, [2nd][DRG], to change the mode to scientific notation. Press the right arrow to underline SCI and press ENTER. This will make the calculator give all of its answers in scientific notation. To change the mode back to normal, underline FLO, "float," again. (You will probably never use ENG, "engineering" notation. All the engineers I know use scientific notation.)
- Use EE, [2nd][x^{-1}], for typing scientific notation. EE means "times 10 to the."
  - 2.34×10^5 is typed 2.34 [2nd][x^{-1}] 5
- For trigonometry, use the [sin], [cos], and [tan] buttons.
  - cos(30) is typed [cos] 30 [ )]
  - $\sin^2(45)$ is typed [sin] 45 [ )] [x^2]
- Use [DRG] to change the mode between degrees and radians when doing trig problems. (You will probably never use GRD, "gradians," in your life. I never have.)
  - tan(π/3) is typed [tan][π][÷] 3 [ )]
  - Don't type [π][A b/c] 3 because when you put a decimal, like π, in a fraction, it confuses the calculator.
- For inverse trig functions, like sin⁻¹, use [2nd][sin], or whichever trig function you need.
  - $\tan^{-1}(\sqrt{3})$ is typed [2nd][tan][2nd][x^2] 3 [ )][ )]

TI-36X Pro

This may be my favorite calculator. If you want to buy a new calculator, and you don't plan to major in engineering in college, this is the calculator I recommend. It can do almost everything a graphing calculator can do, except graph. It is also very user friendly and intuitive.

- One of the most useful but overlooked features is the delete/insert button. When you mistype something, do not clear it. Use the arrow keys to move on top of the mistake, then retype, use [delete] to delete a character, or use insert, [2nd][delete], to insert characters.
- The answer, [2nd][(-)], key, near the [enter] button, can save you from remembering and retyping long numbers.
  - 500 – answer will subtract the previous answer from 500.
- This calculator is prints fractions and mixed numbers intuitively. The [몸] button prints a fraction bar on the screen. You can use the arrow keys to move from top to bottom. To type a mixed number, use 口몸, [2nd] [7]
  - $\frac{4}{5}$ is typed [몸] 4 ► 5
  - $3\frac{1}{2}$ is typed [2nd][7] 3 ► 1 ► 2
- [◄►≈], the button above enter, converts between fraction and decimal, or any exact answer and a decimal.
- To convert between a mixed number and an improper fraction, use [math]. It's the first item in the list, ►n/d◄►Un/d.
- Use the [$x^{\square}$] button for exponents.
  - 2^3 is typed 2 [$x^{\square}$] 3
  - 5^2 is typed 5 [$x^{\square}$] 2 or 5 [x^2]
- Use $^{\square}\sqrt{\ }$, [2nd][$x^{\square}$] for roots.
  - $\sqrt[3]{8}$ is typed 3 [2nd][$x^{\square}$] 8
  - $\sqrt{12}$ is typed 2 [2nd][$x^{\square}$] 12 or [2nd][x^2] 12
  - Again, [◄►≈] converts between the exact answer and a decimal.
- To exit the root, use the right arrow key.
  - $\sqrt{9}+1$ is typed [2nd][x^2] 9 ► + 1

- Use [mode], to change the mode to scientific notation. Press the down arrow then the right arrow to highlight SCI and press [enter]. This will make the calculator give all of its answers in scientific notation. To change the mode back to normal, highlight NORM again. (You will probably never use ENG,"engineering" notation. All the engineers I know use scientific notation.)
- Use [EE] for typing scientific notation. EE means "times 10 to the."
  - 2.34×10^5 is typed 2.34 [EE] 5
- For trigonometry, use the [sin], [cos], and [tan] buttons.
  - cos(30) is typed [cos] 30 [ )]
  - $\sin^2(45)$ is typed [sin] 45 [ )] [x^2]
- Use [mode] to change the mode between degrees and radians when doing trig problems. (You will probably never use GRAD, "gradians," in your life. I never have.)
  - tan(π/3) is typed [tan][$\frac{\square}{\square}$][π]▶ 3 ▶[ )]
- For inverse trig functions, like sin⁻¹, tap [sin] twice.
  - $\tan^{-1}(\sqrt{3})$ is typed [tan][tan][2nd][x^2] 3 ▶[ )]
- For imaginary numbers, change the mode to a+bi. Keep it there always. It won't mess up any other functions.
  - Type $\sqrt{-16}$
- To type in an *i*, press the [π] button three times.
  - 3*i* is typed 3[π][π][π]
  - Type (2 + 3*i*)(4 − 5*i*)
- This calculator also has a unit converter. Press [2nd][8] and look through the options.
  - To convert 6 centimeters to inches, type 6 [2nd][8][enter]▶[enter][enter]
  - Experiment with this. It's really useful!
- num-solv, [2nd][sin], can solve an equation for you. The [*x*] key is beside the [4].
  - Type [2nd][sin] 3 [*x*] + 2 ▶ 17 [enter]. Type a guess (you can leave it at 0) and press [enter] twice.
  - quit, [2nd][mode], exits the program.
  - Experiment with harder equations.
  - If an equation has more than one solution, changing the guess can help you find all of the solutions.
- poly-solv, [2nd][cos], can solve a quadratic or cubic equation for you.
  - To solve $x^2 + 2x - 15 = 0$, type [2nd][cos][enter]. Remember *a*, *b*, and *c* from the quadratic equation. Type 1 [enter] for *a*, 2 [enter] for *b*, and −15 for *c*. Press [enter] to solve. Use the down arrow to see the second solution.
- sys-solv, [2nd][tan], can solve a system of equations for you.
  - To solve $2x + 3y = 13$ and $3x - 4y = -6$, type [2nd][tan][enter]. Fill in the equations on the screen. Type 2[enter] + 3[enter] 13 [enter] 3[enter] − 4[enter] [(-)]6 [enter][enter]

TI-83/TI-84

The TI-83 and TI-84 are essentially the same calculator. The TI-84 has more memory, more speed, and the newer ones have color screens, but functionally, they're the same. It doesn't matter if you have a brand new one or an old hand-me-down. These are wonderful calculators, but I don't recommend students paying for one unless their school requires it or unless they want to major in engineering. If you're going to need it in college, buy it in high school and learn to use it.

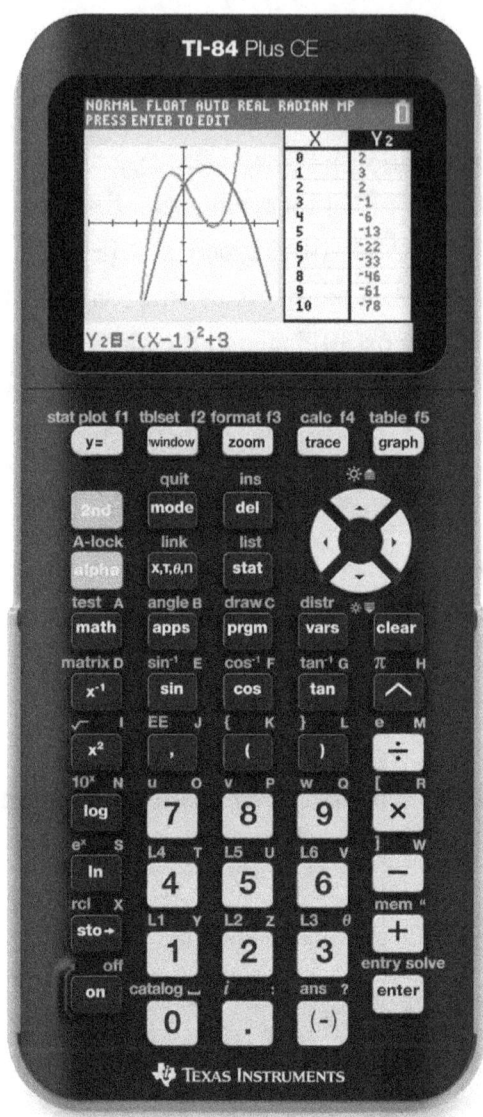

- One of the most useful but overlooked features is the delete/insert button. When you mistype something, do not clear it. Use the arrow keys to move on top of the mistake, then retype, use [del] to delete a character, or use ins, [2nd] [del], to insert characters.
- The answer, or ans, [2nd][(-)], key, near the enter button, can save you from remembering and retyping long numbers.
  - 500 – ans will subtract the previous answer from 500.
- The original TI-83's were lousy at fractions. TI fixed the problem by adding a new menu. Press [alpha][y=] to access it. Option 1, n/d, prints a fraction bar on the screen. You can use the arrow keys to move from top to bottom. To type a mixed number, use option 2, Un/d.
  - $\frac{4}{5}$ is typed [alpha][y=][enter] 4 ► 5
  - $3\frac{1}{2}$ is typed [alpha][y=] ▼ [enter] 3 ► 1 ► 2
- Option 3 on the menu converts between fraction and mixed number.
- To convert a decimal into a fraction, use [math]. It's the first item in the list, ►Frac
- Use the [^] or [x^2] button for exponents.
  - 2^3 is typed 2 [^] 3
  - 5^2 is typed 5 [x^] 2 or 5 [x^2]
- Use √, [2nd][x^2], or ^{x}√, [math] 5, for roots.
  - $\sqrt[3]{8}$ is typed 3 [math] 5 8
  - $\sqrt{16}$ is typed [2nd][x^2] 16

212

- To exit the root, use the right arrow key.
  - $\sqrt{9}+1$ is typed [2nd][x^2] 9 ▶ + 1
- Use [mode], to change the mode to scientific notation. Press the right arrow to highlight SCI and press enter. This will make the calculator give all of its answers in scientific notation. To change the mode back to normal, highlight NORMAL again. (You will probably never use ENG,"engineering" notation. All the engineers I know use scientific notation.)
- Use EE, [2nd][,], for typing scientific notation. EE means "times 10 to the."
  - 2.34×10^5 is typed 2.34 [2nd][,] 5
- For trigonometry, use the [sin], [cos], and [tan] buttons.
  - $\cos(30)$ is typed [cos] 30 [ )]
  - $\sin^2(45)$ is typed [sin] 45 [ )] [x^2]
- Use [mode] to change the mode between degrees and radians when doing trig problems.
  - $\tan(\pi/3)$ is typed [tan][2nd][^][÷] 3 [ )]
- For inverse trig functions, like $\sin^{-1}$, use [2nd][sin].
  - $\tan^{-1}(\sqrt{3})$ is typed [2nd][tan][2nd][x^2] 3 ▶[ )]
- For imaginary numbers, change the mode to a+bi. Keep it there always.
  - Type $\sqrt{-16}$
- To type an i, press [2nd][.].
  - $3i$ is typed 3[2nd][.]
  - Type $(2 + 3i)(4 - 5i)$
- Your calculator may have a unit converter. Go to [apps] and look for SciTools.
  - To convert 6 centimeters to inches, type [apps], go to SciTools,[enter] unit converter, length, 6, highlight cm, [enter], highlight in, [enter].
  - Experiment with this. It's really useful!
  - To exit, use quit, [2nd][mode], repeatedly.
- If there's something you think this calculator should be able to do, but you can't find the command, look under [math]. Explore this menu in your free time. Some of the most useful commands are buried at the bottom of the lists.
- The Solver is located at the bottom of the first list under [math]. It will solve an equation for you. The format varies based on the age of the calculator. It may ask for two expressions (the left and right sides of the equation), or it may want the equation set equal to zero.
  - To solve $x^3 + 6x^2 + 49 = 0$, type [math]▲[enter]. Then type the equation. Use the [X] button beside [alpha]. If you have two boxes, type 0 in the second box. Press [enter].
  - Type a guess (you can leave it at 0). Don't change the "bound=" line.
  - If you press [enter], the calculator won't do anything. You have to press solve, [alpha] [enter] with the cursor on the $x =$ line..
  - If an equation has more than one solution, changing the guess can help you find all of the solutions.
  - To solve a different equation, or when returning to the Solver later, press ▲ to change the equation.

213

- If you're going to use a graphing calculator, you need to know how to graph. To type in the equation, use the [X] button beside [alpha].
  - Press [y=] and type [X][x^2] + [X] – 6 beside the Y_1=. You may have to press the right arrow before you can type.
  - Press [graph]. You should be able to see that the curve crosses the x-axis at –3 and 2. This is a good way to solve simple polynomials.
  - Use [trace] then the left and right arrows to follow the curve. If you graph more than one line, the up and down arrows will switch curves. You can also press [trace] and a number, like –3 or 2, to verify the solutions.
- Graphing can also solve more difficult equations.
  - Go back to [y=], [clear] the first equation, and type in $2x^3 + 5x^2 – 9x – 18$ then [graph].
  - We can see that $x = –3$, $x = 2$, and $x =$ something in between.
  - Go to calc, [2nd][trace], and choose option 2: zero.
  - The calculator will ask for a Left Bound. Use the left arrow to move the cursor to the left of the unknown zero, somewhere around –2. Press [enter]. Then the calculator will ask for a Right Bound. Use the right arrow to move the cursor to the right of the unknown zero, somewhere around –1. Finally, the calculator will ask for a Guess. You can use the arrows to move closer to the zero, or you can just press [enter]. The Zero is X=–1.5.
- Graphing can also solve systems of equations.
  - Go back to [y=], [clear] the previous equation, and type in $Y_1 = 2x – 1$ [enter] and $Y_2 = –3/2x + 6$ [graph].
  - Go to calc, [2nd][trace], and choose option 5: intersect.
  - The calculator will ask for the First Curve. The upper left corner of the screen should say Y1=2X–1. This is the first line, so press [enter]. It will ask for the Second Curve, and it should automatically switch to the second line, so press [enter] again. Finally, the calculator will ask for a Guess. You can use the arrows to move closer to the intersection, or you can just press [enter]. The Intersection is X=2 Y=3.
- There are many other useful commands under calc. Experiment with them.

- If you want to learn some of the most advanced abilities of your calculator, you can program it. If you know the logic of almost any computer programming language, you should be able to pick up TI programming.
- Below, I have included three of my own programs. The first one uses the quadratic formula; the second one uses Cramer's Rule to solve a system of equations; the third one simplifies a square root.
- As of today when I am writing this, these three programs are legal to use on the ACT. Those requirements can change at any time, and it is the student's responsibility to follow the most current regulations.
- Go to [prgm], arrow over to NEW, and press [enter]. Type the name of the new program, press [enter], and copy one of the programs below. To enter the commands, press [prgm] again and select the command from one of the lists.
  - The arrows, →, in the programs are the [sto] key. [sto] saves the result as a variable.

QUAD

```
:ClrHome
:Prompt A,B,C
:B²–4*A*C→D
:(-B+√(D))/(2A)→X
:(-B–√(D))/(2A)→Y
:ClrHome
:Disp "-B+–√(B²–4AC)"
:Disp "        2A"
:Disp X,Y
```

* The spaces on the second-to-last row are [alpha][0].

CRAMER

```
:Disp "AX+BY=C","DX+EY=F"
:Prompt A,B,C,D,E,F
:A*E–B*D→L
:(C*E–F*B)/L→X
:(A*F–D*C)/L→Y
:Disp X,Y
```

SQRT

```
:Prompt X
:1→O
:If X<0
:i→0
:√(abs(X))→A
:int(A)→B
:If (A=B)
:Then
:Disp B*O
:Stop
:End
:While (B≥1)
:abs(X)/B²→D
:int(D)→E
:If (D=E)
:Then
:B*O→O
:D→I
:Goto E
:End
:B–1→B
:End
:Lbl E
:Disp O,"√(",I
:Stop
```

215

1. Mrs. Underwood will choose one member of her 32-student class to represent the class in a quiz tournament. She will choose at random, but the representative can NOT be one of the 3 students with failing grades. What is the probability that Kai, who has a passing grade, will be chosen?

   A. 0

   B. $\dfrac{4}{32}$

   C. $\dfrac{1}{32}$

   D. $\dfrac{1}{3}$

   E. $\dfrac{1}{29}$

2. Big Wave Boat Rentals charges $50 plus $3 an hour to rent a ski boat. Make a Splash Boat Rentals charges $40 plus $5 an hour to rent a similar boat. How many hours must a renter keep a boat to make the total charge for each rental to be equal?

   F. 3

   G. 4

   H. 5

   J. 10

   K. 65

3. When $x = \dfrac{1}{3}$, what is the value of $\dfrac{9x-2}{x}$?

   A. $\dfrac{1}{3}$

   B. 3

   C. $\dfrac{7}{3}$

   D. 7

   E. 15

4. A macaroni and cheese recipe calls for 4 cups of sharp cheddar cheese to make six servings. Judy will modify the recipe by using 6 cups of cheese and proportionally increasing the other ingredients. How many servings will Judy's modified recipe make?

   F. 6

   G. 8

   H. 9

   J. 12

   K. 16

5. For what value of x is the equation $3^{2x+5}=3^{17}$ true?

   A. 3

   B. 6

   C. 9

   D. 27

   E. 33

6. In a certain town, the fine for speeding is $12 for every mile per hour over the posted speed limit. Cameron was fined $216 for speeding on a road with a posted speed limit of 35 mph in that town. At what speed, in mph, was Cameron traveling when he was fined?

   F. 12

   G. 18

   H. 47

   J. 53

   K. 65

7. In the standard (x, y) coordinate plane, what is the midpoint of the line segment connecting the points (5, 7) and (–1, 3)?

   A. (–6, –4)

   B. (–3, –2)

   C. (2, 5)

   D. (4, 10)

   E. (6, 1)

8. The median of 5 distinct numbers is equal to the average of the 5 numbers. The sum of the 5 numbers is 190. What is the sum of the 4 scores that are NOT the median?

F. 120
G. 125
H. 130
J. 152
K. 185

9. A jar contains 9 quarters, 6 dimes, and 12 nickels. If one coin is drawn out at random, what is the probability that a quarter is drawn?

A. $\dfrac{1}{25}$

B. $\dfrac{9}{40}$

C. $\dfrac{1}{9}$

D. $\dfrac{2}{5}$

E. $\dfrac{1}{3}$

10. The graph below shows the fluctuation in temperature on a certain day. Which of the following values gives the positive difference in the greatest temperature and the lowest temperature shown on this graph?

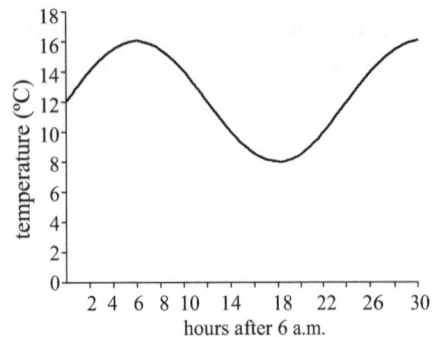

F. 4
G. 8
H. 12
J. 16
K. 24

11. What is the value of the expression below?
$$||2-8|-|-10+5||$$

A. –21
B. –1
C. 1
D. 9
E. 25

12. In parallelogram $ABCD$ below, $\overline{AC}$ is a diagonal, the measure of $\angle ABC$ is 50°, and the measure of $\angle ACD$ is 38°. What is the measure of $\angle CAD$?

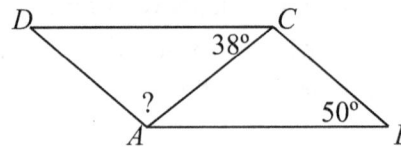

F. 38°
G. 40°
H. 50°
J. 88°
K. 92°

13. What is the sum of the solutions of the 2 equations below?
$$3x+10=25$$
$$4y=14$$

A. 2

B. $8\dfrac{1}{2}$

C. 11

D. 12

E. $15\dfrac{1}{6}$

14. Which of the following expressions is equal to $7^{\frac{3}{4}}$?

F. $\dfrac{7^3}{4}$

G. $\dfrac{7\times 3}{4}$

H. $\sqrt[3]{7}$

J. $\sqrt[3]{7^4}$

K. $\sqrt[4]{7^3}$

15. In the standard (x, y) coordinate plane, what is the slope of the line through the points $(4, -3)$ and $(-1, 3)$?

A. $-\dfrac{6}{5}$

B. $-\dfrac{5}{6}$

C. $\dfrac{5}{6}$

D. $\dfrac{6}{5}$

E. 5

16. Let the function f be defined as $f(x) = 4x^2 - 5(2x + 8)$. What is the value of $f(4)$?

F. -16

G. -6

H. 4

J. 144

K. 176

17. Sean plans to paint the rectangular walls of his room, but he needs to know the surface area of the walls. Two of the walls are 12 feet long, and the other two walls are 14 feet long. The ceiling is 9 feet tall. The total combined area of the window and door is 60 square feet. What is the area, in square feet, that Sean plans to paint?

A. 348

B. 408

C. 428

D. 468

E. 528

18. For which of the following conditions will the sum of integers a and b *always* be an odd integer?

F. a is an odd integer

G. b is an even integer

H. a and b are both even integers

J. a and b are both odd integers

K. a is an even integer and b is an odd integer

19. In $\triangle ABC$ shown below, AB is 21 feet long and BC is 28 feet long. How many feet away from A is the midpoint of $\overline{AC}$?

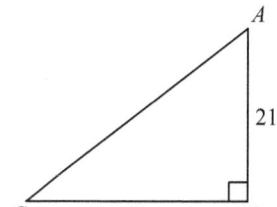

A. 14

B. 17.5

C. 19

D. 24.5

E. 35

20. In $\triangle DEF$, the length of $\overline{DE}$ is $\sqrt{15}$ inches, and the length of $\overline{EF}$ is 4 inches. If it can be determined, what is the length of $\overline{DF}$ in inches?

F. 1 in

G. 4 in

H. $\sqrt{15}$ in

J. $\sqrt{31}$ in

K. Cannot be determined from the given information

21. In the standard (x, y) coordinate plane, what is the slope of the line given by the equation $3x = 5y + 10$?

A. $-\dfrac{3}{5}$

B. $\dfrac{3}{5}$

C. $\dfrac{5}{3}$

D. 3

E. 5

22. The length of a rectangle is 6 inches longer than its width. If the perimeter of the rectangle is 52 inches, what is the width, in inches, of the rectangle?

F. 10

G. 12

H. 16

J. 23

K. 29

23. 6% of 35 is $\frac{1}{10}$ of what number?

A. $\frac{1}{21}$

B. 7

C. 21

D. 210

E. 2,100

24. Enaya is trying to convince his mother to buy him a season pass to the water park. He estimates that he will go to the park 20 times over the summer. Admission to the park for a single day costs $12, and the season pass costs $185. The season pass will admit Enaya into the park at no additional cost as many days as he wants to go. What is the minimum number of days he would have to go to the water park in order for the cost of the season pass to be less than the cost buying a single day admission each time?

F. 9

G. 10

H. 15

J. 16

K. 20

25. $\frac{8.4 \times 10^{-4}}{2.8 \times 10^{-9}} = ?$

A. 3.0×10^{-13}

B. 3.0×10^{-6}

C. 3.0×10^{5}

D. 3.3×10^{5}

E. 3.3×10^{6}

26. The average of a list of 5 numbers is 85. A second list of 5 numbers has the same first 4 numbers, but the fifth number in the original list is 70, and the fifth number in the second list is 105. What is the average of the second list of numbers?

F. 85

G. 89

H. 92

J. 95

K. 102.5

27. Which of the following expressions is a factor of $x^3 + 27$?

A. $x - 3$

B. $x + 3$

C. $x - 27$

D. $x^2 + 9$

E. $x^2 + 3x + 9$

28. A circle in the standard (x, y) coordinate plane has center $O(1, -1)$ and passes through $A(3, 4)$. Line segment $\overline{AB}$ is a diameter of the circle. What are the coordinates of point B?

F. $(4, 3)$

G. $(2, 5)$

H. $(-1, 7)$

J. $(-1, -6)$

K. $(-3, -4)$

29. The number n is located at -1.5 on the number line below.

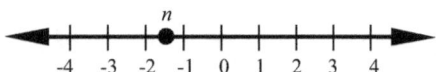

One of the following number lines shows the location of n^2. Which number line is it?

A.

B.

C.

D.

E.

30. The number 1,001 is the product of the prime numbers 7, 11, and 13. Knowing this, what is the prime factorization of 50,050?

F. $5 \cdot 7 \cdot 11 \cdot 13$

G. $2 \cdot 5 \cdot 7 \cdot 11 \cdot 13$

H. $50 \cdot 7 \cdot 11 \cdot 13$

J. $5 \cdot 10 \cdot 7 \cdot 11 \cdot 13$

K. $2 \cdot 5^2 \cdot 7 \cdot 11 \cdot 13$

31. Katrina made an apple pie, but she only ate $\frac{2}{7}$ of it. She gave the remaining pie to her three little brothers. What fraction of the whole pie will each of Katrina's brothers eat if they share the remaining pie equally?

A. $\frac{5}{7}$

B. $\frac{3}{5}$

C. $\frac{1}{3}$

D. $\frac{5}{21}$

E. $\frac{2}{21}$

Use the following information to answer questions 32 – 34.

Lilly Kae proposes a new trapezoidal park to the Redrock Parks and Recreation department. The figure below shows the scale drawing of her proposal, where one inch represents 2.5 feet. It shows the sides and dimensions of the equipment in inches.

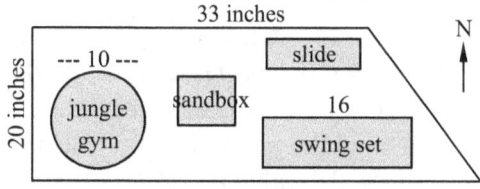

32. What is the area, in square inches, of the scale drawing of the park?

F. 660
G. 810
H. 960
J. 975
K. 1620

33. Lilly Kae proposes installing a fence around the park. What is the park's perimeter, in *feet*?

A. 101
B. 126
C. 252.5
D. 278
E. 315

34. The length of the south side of the park is what percent of the length of the north side of the park?

F. 115%
G. 122%
H. 131%
J. 140.5%
K. $145 \frac{5}{11}$ %

35. Which of the following expressions, when evaluated, equals an irrational number?

A. $\frac{\sqrt{3}}{\sqrt{12}}$

B. $\frac{\sqrt{18}}{\sqrt{2}}$

C. $\left(\sqrt{6}\right)^2$

D. $\sqrt{3} \times \sqrt{12}$

E. $\sqrt{18} + \sqrt{2}$

36. The equation $|2x+5|+4=8$ has 2 solutions. Those solutions are equal to the solutions of which of the following pairs of equations?

F. $\begin{matrix} 2x+9= 8 \\ -2x+9=-8 \end{matrix}$

G. $\begin{matrix} 2x+5= 4 \\ -2x+5= 4 \end{matrix}$

H. $\begin{matrix} 2x+5= 12 \\ -(2x+5)=12 \end{matrix}$

J. $\begin{matrix} 2x+5= 4 \\ -(2x+5)=12 \end{matrix}$

K. $\begin{matrix} 2x+5= 4 \\ -(2x+5)=4 \end{matrix}$

37. The graph of $y=(x-3)^2$ is in the standard (x,y) coordinate plane. Which of the following transformations, when applied to the graph of $y=x^2$, results in the graph of $y=(x-3)^2$?

A. Translation to the right 3 units
B. Translation to the left 3 units
C. Translation up 3 units
D. Translation down 3 units
E. Reflection across the line $x = 3$

The Davis family is planning to build an addition on their home which consists of a bedroom (BR) and two closets (C). Shown below are the floor plan (left figure) and a side view of the addition (right figure). In the side view, the roof forms an isosceles triangle ($\triangle ABC$), the walls are perpendicular to the level floor ($\overline{ED}$), $\overline{AC} \parallel \overline{ED}$, F is the midpoint of $\overline{AC}$, and $\overline{BF} \perp \overline{AC}$.

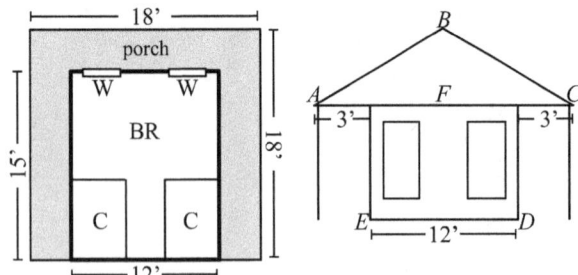

During the week that the addition's roof will be built, there is a 30% chance of rain each day.

38. Mrs. Davis wants a 3-foot-wide porch built around three sides of the addition, as shown in the floor plan. What will be the area, in square feet, of the porch?
F. 54
G. 63
H. 144
J. 162
K. 324

39. Mrs. Davis will install a light fixture in each closet and a ceiling fan in the bedroom. Then she will hang curtains on the two windows. The cost of the curtains is $42.75 per window. The cost of a light fixture is $35.95, and the cost of a ceiling fan is 175% the cost of a light fixture. How much will Mrs. Davis will pay for curtains, light fixtures, and ceiling fans?
A. $141.61
B. $157.40
C. $211.33
D. $220.31
E. $283.23

40. Mr. Davis plans to roof the addition 2 consecutive days. Assuming that the chance of rain is independent of the day, what is the probability that it will rain both days?
F. 0.09
G. 0.15
H. 0.21
J. 0.30
K. 0.60

41. The number of decibels, d, produced by an audio source can be modeled by the equation $d = 10 \log\left(\dfrac{I}{K}\right)$, where I is the sound intensity of the audio source and K is a constant. How many decibels are produced by an audio source whose sound intensity is 10,000 times the value of K?

A.   4
B.   40
C.   400
D.   10,000
E.   100,000

42. The line in the standard (x,y) coordinate plane shown below passes through the origin and $(8, 2)$. The acute angle between the line and the positive x-axis has measure θ. What is the value of $\tan \theta$?

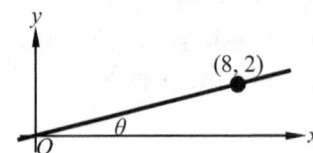

F. $\dfrac{\sqrt{17}}{4}$

G. $\dfrac{1}{\sqrt{17}}$

H. $\dfrac{4}{\sqrt{17}}$

J. $\dfrac{1}{4}$

K. $\dfrac{4}{1}$

43. The frequency chart below shows the cumulative number of soccer players on a certain team whose season total goals fell within certain ranges.

Score range	Cumulative number of players
1 – 4	6
1 – 8	9
1 – 12	11
1 – 15	12

How many players scored between 9 and 12 goals, inclusive?

A. 2
B. 4
C. 6
D. 9
E. 11

44. Jayden wants to find the volume of his solid toy action figure. He fills a rectangular plastic box 6 cm wide, 10 cm long, and 12 cm high with water to a depth of 5 cm. Jayden submerged the action figure under water. The water level rose to a height of 8.2 cm. Which of the following is closest to the volume, in cubic centimeters, of the toy action figure?

F. 192
G. 230
H. 300
J. 492
K. 720

45. Amazon is shipping a cylindrical container of popcorn with a height of 7.25 inches and a diameter of 6.5 inches. The tin ships in a cubical box with an interior side length of 8 inches. The interior of the box not occupied by the tin is filled with packing peanuts. Which of the following expressions gives the number of cubic inches of the box filled with packing peanuts?

A. $6(8)^2 - 2\pi(7.25)(6.5) - 2\pi(3.25)^2$
B. $6(8)^2 - 2\pi(7.25)(6.5)$
C. $8^3 - \pi(7.25)(6.5)^2$
D. $8^3 - \pi(7.25)(3.25)^2$
E. $8^3 - \pi(6.5)^3$

46. The graphs below show the costs of parking in two separate parking garages. Garage A Charges a whole dollar amount for every 2 hours parked. Garage B charges a flat rate for the first hour plus a smaller whole dollar amount for every 2 hours after the first. When the two garages are compared, what is the cheaper cost to park in a garage for 4 hours?

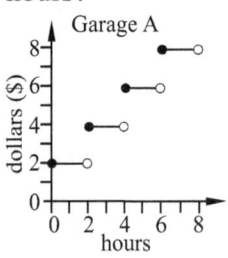 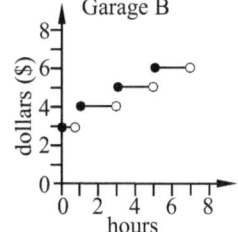

F. $3
G. $4
H. $5
J. $6
K. $8

47. The graph of a function $y = f(x)$ consists of 3 line segments. The graph below shows the line segments and the coordinates of the endpoints in the standard (x, y) coordinate plane. What is the area, in square coordinate units, of the region bounded by the graph of $y = f(x)$, the positive x-axis, and the positive y-axis?

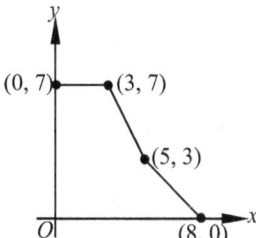

A. 24
B. 28
C. 35.5
D. 37.5
E. 56

48. A rectangle has dimensions of 9 feet by 24 feet. What is its area in square *yards*?

F. 24
G. 48
H. 72
J. 108
K. 216

49. The table below shows Kobe Bryant's scoring statistics for a recent NBA game. How many points did Kobe score in the game?

Type of shot	Number attempted	Percent successful
1-point free throw	20	90%
2-point field goal	33	64%
3-point field goal	13	54%

- **A.** 46
- **B.** 66
- **C.** 75
- **D.** 81
- **E.** 125

50. The sum of two positive integers is 160. The smaller number is equal to 4 more than the square root of the larger number. What is the difference of the two numbers?
- **F.** 16
- **G.** 78
- **H.** 82
- **J.** 128
- **K.** 144

51. In the following system of equations, a, b, and c are real numbers.
$$mx + ny = p$$
$$y = x^2$$
For which of the following conditions will the system have more than one (x,y) solution with real number coordinates?
- **A.** $m^2 + 4np > 0$
- **B.** $m^2 - 4np < 0$
- **C.** $n^2 - 4mp > 0$
- **D.** $n^2 - 4mp < 0$
- **E.** $n^2 + 4mp > 0$

52. If the 4th term of an arithmetic sequence is 8 and the 7th term is 19, what is the 31st term of the sequence?
- **F.** 107
- **G.** 118
- **H.** 121
- **J.** 124
- **K.** 129

53. For what positive real value of n does the determinant of the matrix $\begin{bmatrix} n & 2 \\ 3 & n \end{bmatrix}$ equal n?

(Note: The determinant of matrix $\begin{bmatrix} a & b \\ c & d \end{bmatrix}$ equals $ad - bc$.)
- **A.** 2
- **B.** 3
- **C.** $\sqrt{6}$
- **D.** 6
- **E.** No such value of n exists.

54. What is period of the function $y = \sec\left(\dfrac{x}{2}\right)$?
- **F.** $\dfrac{\pi}{4}$
- **G.** $\dfrac{\pi}{2}$
- **H.** π
- **J.** 2π
- **K.** 4π

55. One of the following graphs in the standard (x,y) coordinate plane shows the graph of $y = \sin^2 x + \cos^2 x$ over the domain $-\pi \leq x \leq \pi$. Which one?

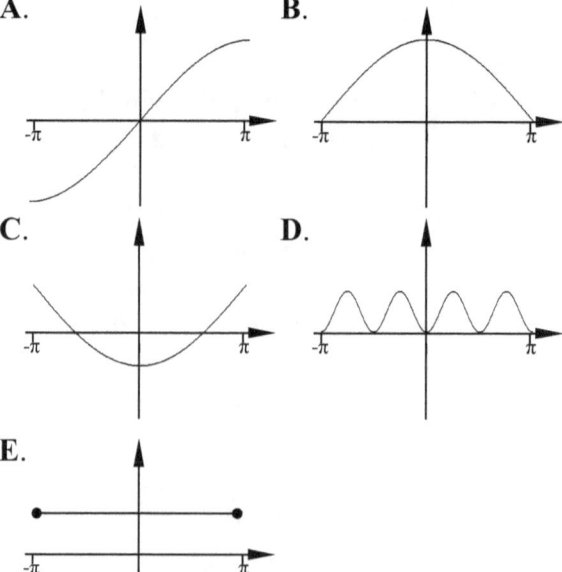

56. Toby and Kevin are playing a game in the school cafeteria. Each boy flips three pennies on his turn. For each coin that lands on heads, he gets five points. Let the random variable X represent the total points awarded on any turn of the game. What is the expected value of X?

F. $\dfrac{1}{2}$

G. $\dfrac{5}{2}$

H. 5

J. $\dfrac{15}{2}$

K. 15

57. Ray $\overrightarrow{BD}$ bisects $\angle ABC$, the measure of $\angle ABD$ is $(2x+10)°$ and the measure of $\angle ABC$ is $6x°$. What is the measure of $\angle CBD$?

A. $10°$
B. $15°$
C. $30°$
D. $60°$
E. $120°$

58. Over the domain of $0\leq\theta\leq\pi$, the inequality $|\cos\theta|\geq1$ is true for which of the following values of θ?

F. $\{0,\pi\}$

G. $\{0\}$

H. $\left\{\theta\,|\,0\leq\theta\leq\dfrac{\pi}{2}\right\}$

J. $\{\theta\,|\,0\leq\theta\leq\pi\}$

K. $\{\varnothing\}$

59. The set of numbers {5, 8, 11, 12, A, B, 22, 30} has a mode of 30 and a median of 15. What is the mean of the set of numbers?

A. 14
B. 17
C. 19
D. 22
E. 30

60. Given a positive integer k such that $i^k=1$, which of the following is true of the remainder when k is divided by 4?

F. The remainder is 0.
G. The remainder is 1.
H. The remainder is 2.
J. The remainder is 3.
K. The remainder is greater than 4.

Sample Test 1 Answers and Explanations

1. **E.** $\frac{1}{29}$ 3 students cannot be chosen, so that leaves 29 who can. Kai is 1 out of those 29, so the probability is $\frac{1}{29}$.

2. **H.** 5 Big Wave's cost is $50+3x$ and Make a Splash's cost is $40+5x$. Set them equal and solve. $50+3x=40+5x$ $x=5$

3. **B.** 3 Plug in $\frac{9(1/3)-2}{1/3}=\frac{3-2}{1/3}=\frac{1}{1/3}$ $=1\times3/1=3$

4. **H.** 9 Set up ratios $\frac{\text{cheese}}{\text{servings}}=\frac{4}{6}=\frac{6}{x}$ and cross-multiply $4x=36$ so $x=9$

5. **B.** 6 $3^{2x+5}=3^{17}$ Since the bases are equal, the exponents must be equal. $2x+5=17$ Solve. $x=6$

6. **J.** 53 $12x=216$ so $x=18$. That means Cameron was driving 18 mph over the speed limit of 35 mph. $35+18=53$

7. **C.** (2, 5) To find the midpoint, average the x's and average the y's. $x=\frac{5-1}{2}=2$ and $y=\frac{7+3}{2}=5$ so the midpoint is (2, 5).

8. **J.** 152 The average of 5 numbers is the sum divided by 5. So the average is $\frac{190}{5}=38$ which is also the median (the number in the middle). Make up an easy set of 5 numbers: 36, 37, 38, 39, 40. The median and the average of this set equal 38. The sum of the other four numbers is $36+37+39+40=152$

9. **E.** $\frac{1}{3}$ $\frac{9 \text{ quarters}}{27 \text{ coins}}=\frac{1}{3}$

10. **G.** 8 16 (greatest) minus 8 (lowest) = 8

11. **C.** 1 Absolute values act like parentheses, so we work what's inside them first. $||2-8|-|-10+5||=||-6|-|-5||$ Absolute values make numbers positive, so we have $|6-5|=|1|=1$

12. **K.** 92° Using alternate interior angles and properties of parallelograms, we can label 2 more angles. The three angles in a triangle add up to 180°, so $50+38+x=180$ and $x=92°$

13. **B.** $8\frac{1}{2}$ This question looks harder than it really is. Solve the two equations separately. $3x+10=25$ $3x=15$ $x=5$ $4y=14$ $y=3.5$ The question asked for the sum. $5+3.5=8.5=8\frac{1}{2}$

14. **K.** $\sqrt[4]{7^3}$ The top of a fraction exponent is a normal exponent. The bottom is the root. $7^{\frac{3}{4}}=\sqrt[4]{7^3}$

15. **A.** $-\frac{6}{5}$ $\text{slope}=\frac{\text{rise}}{\text{run}}=\frac{3-(-3)}{-1-4}=-\frac{6}{5}$

16. **F.** -16 Plug in $f(4)=4(4)^2-5(2(4)+8)$ and simplify $4(16)-5(8+8)=64-5(16)$ $64-80=-16$

17. **B.** 408 Area is *base* times *height*. Two walls are $12\times9=108$ and the other two are $14\times9=126$. Add the four walls and subtract the windows and door. $2(108)+2(126)-60=408$

18. **K.** a is an even number and b is an odd number Test the answer choices using easy odd and even numbers, like 1 and 2. Which one always has an odd sum? $1+2=3$

19. **B.** 17.5 Find the length of $\overline{AC}$ using $a^2+b^2=c^2$. $21^2+28^2=c^2$ $1225=c^2$ $c=\sqrt{1225}=35$ That means A is 35 feet from C. The midpoint is half of 35, or 17.5.

20. **K.** Cannot be determined from the given information. You don't know the shape of the triangle. It could be either of these, or anywhere in between.

226

21. B. $\frac{3}{5}$ Rearrange the equation to get $y=mx+b$ form. $y=\frac{3}{5}x-2$ m is the slope.

22. F. 10 Perimeter adds up all the sides. If the width is x and the length is $x+6$, the perimeter is $x+(x+6)+x+(x+6)=52$. Solve. $4x+12=52$ $x=10$

23. C. 21 In a percent problem, "of" means times and "is" means equals.

$6\%\times 35=\frac{1}{10}\times n$ $.06\times 35=\frac{n}{10}$

$2.1=\frac{n}{10}$ $21=n$

24. J. 16 Set up the problem. $12x=185$ $x=15.4$ Since Enaya can't go 15.4 times, he has to go more than 15 times; choose 16.

25. C. 3.0×10^{5} Use a calculator first. Be careful with parentheses. $\frac{8.4\times 10^{-4}}{2.8\times 10^{-9}}$ Can be entered as $(8.4\times 10^{-4})/(2.8\times 10^{-9})$. The calculator gives 300000. 3 with 5 zeros is 3×10^{5}.

26. H. 92 Make up numbers. If the average is 85, I'd like the numbers to be 85, 85, 85, 85, and 85. The problem says the 5th number is 70. That's smaller by 15, so we need to make another number bigger by 15 to keep the same average. My list is now 85, 85, 85, 100, and 70. Now I know the second list is 85, 85, 85, 100, and 105. The average is $\frac{85+85+85+100+105}{5}=92$.

27. B. $x+3$ $x^{3}+27$ factors into $(x+3)(x^{2}-3x+9)$. Only the first parentheses is an answer choice.

28. J. $(-1, -6)$ First work the x's. Find the pattern. To get from A to O you go from 3 to 1 by subtracting 2. To get from O to B, subtract 2 again. The x's are 3, 1, –1. The y's are 4, –1, ? The pattern is subtract 5. 4, –1, –6.

$\underset{A}{\overset{3,4}{\rule{0pt}{0pt}}} \rule{2cm}{0.4pt} \underset{O}{\overset{1,-1}{\rule{0pt}{0pt}}} \rule{2cm}{0.4pt} \underset{B}{\overset{-1,-6}{\rule{0pt}{0pt}}}$

29. E.

$(-1.5)^{2}=2.25$ A negative squared is positive.

30. K. $2\cdot 5^{2}\cdot 7\cdot 11\cdot 13$ $50,050=1,001\times 50$ The prime factorization of 50 is $2\cdot 5^{2}$ and we are told that the prime factorization of 1,001 is $2\cdot 5^{2}\cdot 7\cdot 11\cdot 13$. List all factors.

31. D. $\frac{5}{21}$ 1 whole pie $-\frac{2}{7}=\frac{5}{7}$ remaining.

Divide that by 3. $\frac{5}{7}\cdot\frac{1}{3}=\frac{5}{21}$

32. G. 810 The area of a trapezoid is $A=\frac{1}{2}(b_{1}+b_{2})h=\frac{1}{2}(48+33)\cdot 20=810$

33. E. 315 To find the diagonal side, break the trapezoid into a rectangle and a triangle. Use the Pythagorean theorem. The perimeter is $33+20+48+25=126$ inches but the question asked for *feet*. The scale is 1 inch to 2.5 feet. $126\times 2.5=315$ *feet*

34. K. $145\frac{5}{11}\%$ $48=p\times 33$ $p=1.\overline{45}$

35. E. $\sqrt{18}+\sqrt{2}$ An irrational number can't be simplified into a fraction. $\sqrt{18}+\sqrt{2}=4\sqrt{2}$

36. K. $\begin{array}{l}2x+5=\ \ 4\\ -(2x+5)=4\end{array}$ First solve for the absolute value. $|2x+5|=4$ To split it into two equations, write it once without the absolute value and once making one side negative.

37. A. Translation to the right 3 units The equation $y-k=f(x-h)$ shifts the point $(0,0)$ to (h,k). Our equation can be written $y-0=f(x-3)$ and (h,k) is $(3,0)$.

38. H. 144 The area of the porch is the big rectangle minus the small rectangle. $A=l\times w=18\times 18-12\times 15=324-180=144$

39. D. $220.31 We need two light fixtures at $35.95 each, plus a ceiling fan at $1.75\times 35.95=\$62.91$, and two sets of curtains at $42.75 each. $2(35.95)+62.91+2(42.75)=\$220.31$

40. F. 0.09 The probability of rain each day is 30%, or 0.30. To find the probability it rains both days, we multiply $0.3\times 0.3=0.09$

41. B. 40 Plug in $I=10,000K$. $d=10\log\left(\frac{10,000K}{K}\right)=10\log(10,000)$

Use a calculator. $10\log(10,000)=40$

42. J. $\frac{1}{4}$ Tangent equals opposite over adjacent. We know the point (8, 2) is on the line, so the x side (adjacent) is 8 and the y side (opposite) is 2. $\tan\theta=\frac{8}{2}=\frac{1}{4}$

43. A. 2   11 people scored between 1 and 12 goals, but 9 of them scored 8 or fewer goals. $11 - 9 = 2$

44. F. 192   The volume of the toy equals the volume of the water with the toy minus the water alone. $V=lwh$, so the volume of the water alone was $V=10\cdot6\cdot5=300$ and the volume of the water with the toy was $V=10\cdot6\cdot8.2=492$. The volume of the toy is $492 - 300 = 192$.

45. D. $8^3-\pi(7.25)(3.25)^2$ The volume of the packing peanuts equals the volume of the box minus the volume of the tin. The box is $V=lwh=8^3$ and the volume of the tin is $V=\pi r^2 h=\pi(3.25)^2(7.25)$ Subtract.

46. H. \$5   The cost for Garage A is \$6 because we have to use the solid point at the 6 instead of the open circle at the 4. The cost for Garage B is \$5, which is cheaper.

47. C. 35.5   Cut the shape into two trapezoids or two rectangles and two triangles. The trapezoids are $\frac{1}{2}(b_1+b_2)h=\frac{1}{2}(3+5)4=16$ and $\frac{1}{2}(b_1+b_2)h=\frac{1}{2}(5+8)3=19.5$. Add $16+19.5=35.5$

48. F. 24   First, convert to yards. Remember that 3 feet equals 1 yard. 9 feet equals 3 yards and 24 feet equals 8 yards. Now the area is $3\times8=24$ square yards.

49. D. 81   For each type of shot, we need to know how many he made. Multiply the number attempted by the percent successful and then multiply by the points. Free throws are $20\times0.90\times1=18$, 2-points are $33\times0.64\times2=42$, and 3-points are $13\times0.54\times3=21$. The total is 81.

50. J. 128   Write the two equations. The sum of two integers is 160 becomes $x+y=160$. The smaller is 4 more than the square root of the larger becomes $y=\sqrt{x}+4$. Substitute to get $x+\sqrt{x}+4=160$. Simplify.

$\sqrt{x}=156-x$ You can square and factor, but the numbers will be big. You can also guess and check. Choose perfect squares because of the square root. If $x=100$, $10=156-100=56$. Guess bigger. If $x=144$, $12=156-144=12$. It works. So $x = 144$ and $y = 16$. The problem asked for the difference. $144 - 16 = 128$.

51. A. $m^2+4np>0$ Substitute $y=x^2$ into the first equation. $mx+nx^2=p$. Rearrange to standard form. $nx^2+mx-p=0$ Use the quadratic formula. $\dfrac{-m\pm\sqrt{m^2-4(n)(-p)}}{2n}$ To get two real solutions, the determinant (the thing in the square root) must be positive. Just cancel the two negatives. $m^2+4np>0$

52. F. 107   The common difference is $19 - 8 = 11$ divided by $7 - 4 = 3$. So $d=11/3$. To get from the 7th term to the 31st, we need 24 more terms. $19+24(11/3)=107$

53. B. 3   The problem gives the formula for the determinant. Plug in $n(n)-2(3)=n$ and simplify. $n^2-n-6=0$ Factor. $(n-3)(n+2)=0$ So x equals 3 or –2. The problem asked for the positive value, 3.

54. K. 4π The period of secant normally is 2π, so set $\dfrac{x}{2}=2\pi$ and solve. $x = 4\pi$.

55. E. $\sin^2x+\cos^2x$ is an identity which always equals 1. Try plugging in different values for x. Since it's a constant, choose **E**.

56. J. $\dfrac{15}{2}$ Each coin gives 5 points half of the time, and there are three coin flips per turn. $\dfrac{5}{2}\times3=\dfrac{15}{2}$

57. C. 30°   Draw a picture. Bisects means cut in half. $2(2x+10)=6x$ Solve. $x=10$ The problem asked for ∠CBD so $(2(10)+10)=30$

58. F. $\{0,\pi\}$ Cosine can never be greater than 1, so we only have to find where it equals 1. $\cos(0)=1$ and $\cos(\pi)=1$.

59. B. 17   We have to figure out what A and B equal. Since the mode is 30, we know at

228

least one of them is 30. If we leave out the other unknown, the median is 12, so we know the other unknown is greater than 12. Now our list is 5, 8, 11, 12, A, 22, 30, 30. Since there are an even number of terms, the median is the average of the middle two. 12 and what give an average of 15? A has to equal 18 because $\frac{12+18}{2}=15$. The mean is

$$\frac{5+8+11+12+18+22+30+30}{8}=17$$

60. F. The remainder is 0   The imaginary number $i=\sqrt{-1}$. Find a number k such that $i^k=1$ the simplest value for k is 4. $i^2=-1$ so $i^4=(-1)^2=1$. When 4 is divided by 4 the remainder is 0.

Sample Test 2

1. The length of a rectangle is 18 inches. The width of the rectangle is $\frac{1}{2}$ the length. What is the perimeter of the rectangle, in inches?

   A. 27
   B. 36
   C. 48
   D. 54
   E. 162

2. What is the value of $|3-x|$ when $x = 11$?
   F. −14
   G. −8
   H. 8
   J. 11
   K. 14

3. $(8p - 5q) - (4q + 6p)$ is equivalent to:
   A. $2p - 9q$
   B. $2p - q$
   C. $3p + 2q$
   D. $4p - 11q$
   E. $4p + q$

4. At a craft store, Karyn spent $3.05 on red, blue, and yellow beads. Red beads cost $0.15 each, blue beads cost $0.20, and yellow beads cost $0.15 each. Karyn bought 8 red beads and 4 blue beads. How many yellow beads did she buy?
   F. 4
   G. 6
   H. 7
   J. 9
   K. 18

5. Each week, Louisa will be paid $110 plus 15% of her sales. Let s represent Louisa's total sales for the next week. Which of the following is an expression representing Louisa's pay, in dollars, for the hours she is scheduled to work next week?
   A. $0.15s + 110$
   B. $0.15s + 1.10$
   C. $1.10s + 0.15$
   D. $15s + 110$
   E. $110s + 0.15$

6. Which of the following numbers has the greatest value?
   F. $0.\overline{5}$
   G. 0.5
   H. 0.55
   J. 0.555
   K. 0.5555

7. Every morning, Jared monitored the water level of a river before and during a flood. On the 1st day, the river was 6 inches below flood stage. On the 5th day, the river was 16 inches above flood stage. What was the change in the water level of the river from the 1st day to the 5th day?
   A. −22 in.
   B. −10 in.
   C. 6 in.
   D. 10 in.
   E. 22 in.

8. If $f(x) = 3x^2 + 2x - 5$, then $f(-2) = ?$
   F. −21
   G. −13
   H. 3
   J. 11
   K. 13

9. The figure below shows lines *l* and *m*, line segments $\overline{AC}$ and $\overline{AD}$, and the measures of two angles. What is the measure of $\angle CAD$?

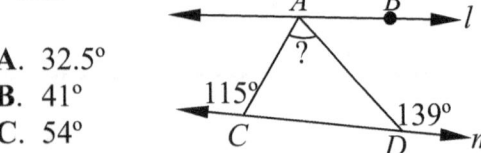

- **A.** 32.5°
- **B.** 41°
- **C.** 54°
- **D.** 65°
- **E.** 74°

10. What is the median of the list of numbers below?

$$8, 5, 6, 10, 9, 6, 12$$

- **F.** 5
- **G.** 6
- **H.** 8
- **J.** 9
- **K.** 10

11. The population of Actville grows exponentially and can be modeled by the equation $P(t) = 25,000(1.15)^t$, where *t* is the number of years after January 2018. According to this model, which of the following numbers is closest to the projected population of Actville in January 2021?

- **A.** 29,000
- **B.** 33,000
- **C.** 38,000
- **D.** 44,000
- **E.** 47,000

12. The dimensions of the rectangle below are given in millimeters. Which of the following expressions gives the area, in mm^2, of the rectangle?

$x + 5$

$2x - 3$

- **F.** $6x + 4$
- **G.** $x^2 + 7x - 15$
- **H.** $2x^2 - 15$
- **J.** $2x^2 + 7x - 15$
- **K.** $2x^2 + 13x + 15$

13. Graphs A, B, and C below each represent the movement of an object as a function of time. In each graph, time is plotted on the *x*-axis.

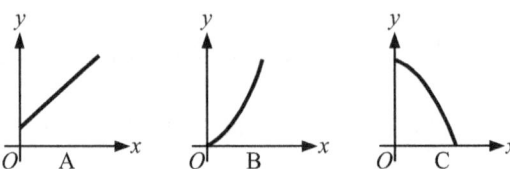

The phrases below each represent the movement plotted on one of the graphs above. Which phrase describes which graph?

I. The distance a train travels while accelerating.

II. The height of water in a graduated cylinder filled at a constant rate.

III. The height of a ball falling toward the ground.

	Graph A	Graph B	Graph C
A.	I	II	III
B.	II	I	III
C.	II	III	I
D.	III	I	II
E.	III	II	I

14. Which of the following augmented matrices represents the system of linear equations below?

$$2x + 3y = 4$$
$$5x - y = 23$$

- **F.** $\begin{bmatrix} 2 & 3 & -4 \\ 5 & -1 & -23 \end{bmatrix}$
- **G.** $\begin{bmatrix} 2 & 3 & 4 \\ 5 & -1 & 23 \end{bmatrix}$
- **H.** $\begin{bmatrix} 2 & 3 & 4 \\ 5 & 0 & 23 \end{bmatrix}$
- **J.** $\begin{bmatrix} 2 & 3 & 4 \\ 5 & 1 & 23 \end{bmatrix}$
- **K.** $\begin{bmatrix} 2 & 5 & 4 \\ 3 & -1 & 23 \end{bmatrix}$

15. Je'Mari earned grades of 80, 75, 90, and 87 on his first four Literature essays. He has one more essay to write. What is the minimum grade Je'Mari needs to earn on the last essay so that his mean grade on all five essays is two points higher than his mean grade on the first four essays?

A. 81
B. 83
C. 85
D. 93
E. 95

16. In the standard (x,y) coordinate plane below, 6 points are labeled on a parabola. Which of the following lines has the slope of *least* value?

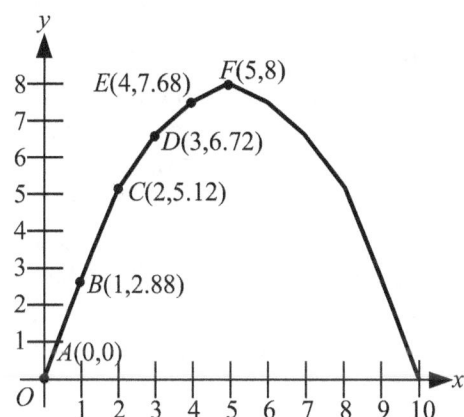

F. $\overline{AB}$
G. $\overline{BC}$
H. $\overline{CD}$
J. $\overline{DE}$
K. $\overline{EF}$

17. Aiden purchased a used pickup truck for $11,500, including all costs and taxes. He paid $1,500 as a down payment and signed a loan for the remainder of the price. To pay off the loan, he paid 60 payments of $275. The total of all Aiden's payments, including the down payment, was how much more than the purchase price of the truck?

A. $ 500
B. $ 6,500
C. $ 8,500
D. $15,000
E. $16,500

18. A sample of 50 valves consisting of two types, ball valves and butterfly valves, and made by two companies, ACME Valve or Bruce's Valve Co., were found to be defective. The table below shows the type and manufacturer of the defective valves.

Type of valve	Number of valves made by company:	
	ACME	Bruce's
ball	12	21
butterfly	8	9

What is the probability that a randomly selected valve from this group is a ball valve made by ACME Valve?

F. $\dfrac{6}{25}$

G. $\dfrac{21}{50}$

H. $\dfrac{12}{33}$

J. $\dfrac{4}{25}$

K. $\dfrac{9}{50}$

19. Builders describe the slope of a roof using the term *pitch*. For example, a roof with a pitch of $\dfrac{1}{3}$ has a vertical rise of 1 foot for every horizontal run of 3 feet. The figure below shows a 25-foot-long roof with a vertical rise of y feet and a horizontal distance of 20 feet. What is the pitch of this roof?

A. $\dfrac{1}{2}$

B. $\dfrac{1}{4}$

C. $\dfrac{3}{4}$

D. $\dfrac{3}{5}$

E. $\dfrac{4}{5}$

20. Given that $x \geq 3$ and $x + y \leq 8$, what is the GREATEST value that y can have?

F. –3
G. 2
H. 4
J. 5
K. 11

21. In the figure below, all of the small squares are equal in area, and the area of rectangle *ABCD* is 1 square foot. Which of the following expressions represents the area, in square feet, of the shaded region?

A. $\dfrac{1}{8} \cdot \dfrac{1}{5}$

B. $\dfrac{1}{8} \cdot \dfrac{4}{5}$

C. $\dfrac{1}{8} \cdot \dfrac{7}{8}$

D. $\dfrac{7}{8} \cdot \dfrac{1}{5}$

E. $\dfrac{7}{8} \cdot \dfrac{4}{5}$

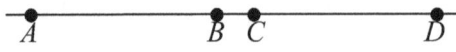

22. Sophie bought a new television that was on sale for 20% less than the sticker price of $1,500. She also paid a sales tax equal to 8% of the discounted price. What is the total price Sophie paid for her new television?

F. $1,104
G. $1,200
H. $1,228
J. $1,296
K. $1,320

23. As shown below, points *A*, *B*, *C*, and *D* are collinear. Point *B* is between *A* and *C*, and point *C* is between *B* and *D*. Given that $AC = BD = 10$ inches and $BC = 2$ inches, what is *AD*, in inches?

A. 8
B. 12
C. 18
D. 20
E. 22

24. Lexington Ave. and Spencer Ave. are parallel to each other. Kentwood Street intersects Lexington Ave. to form a 55° angle, as shown below. What is the measure of the angle formed at the northeast corner of Spencer Ave. and Kentwood Street?

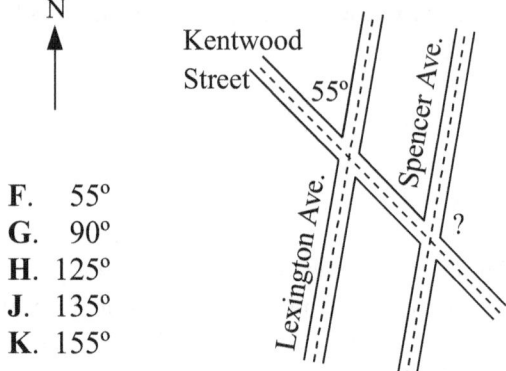

F. 55°
G. 90°
H. 125°
J. 135°
K. 155°

25. On the first day of their camp-out, the Cub Scouts set up camp in the Kisatchie National Forest. The next morning, they hiked to a picnic area 4 miles due south and 3 miles due west of their campsite. If the elevation of the picnic area is the same as the elevation of the campsite, what is the straight-line distance from the campsite to the picnic area, to the nearest 0.1 mile?

A. 3.5
B. 3.7
C. 5.0
D. 5.5
E. 7.0

26. Which of the following equations represents the line in the standard (x,y) coordinate plane that passes through $(3, -2)$ and has a slope of $-\dfrac{1}{3}$?

F. $y = -3x + 1$

G. $y = -\dfrac{1}{3}x - 1$

H. $y = -\dfrac{1}{3}x + 3$

J. $y = \dfrac{1}{3}x - 1$

K. $y = 3x - 11$

27. A number, n, when divided by its reciprocal, has a result of $\dfrac{25}{9}$. What is the value of n?

A. $\dfrac{5}{3}$

B. $\dfrac{5}{9}$

C. $\dfrac{3}{5}$

D. $\dfrac{25}{27}$

E. $\dfrac{27}{5}$

28. The figure below shows a ladder leaning against a house. The base of the ladder is placed 5 feet from the house, and the length of the ladder is 13 feet. Which of the following expressions is closest to the angle of elevation between the ladder and the horizontal ground?

F. $\sin^{-1}\left(\dfrac{5}{13}\right)$

G. $\sin^{-1}\left(\dfrac{13}{5}\right)$

H. $\cos^{-1}\left(\dfrac{5}{13}\right)$

J. $\cos^{-1}\left(\dfrac{13}{5}\right)$

K. $\tan^{-1}\left(\dfrac{5}{13}\right)$

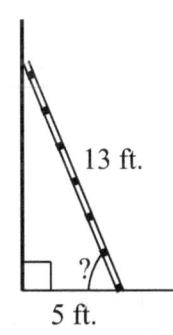

13 ft.

?

5 ft.

29. Of the 700 students enrolled at Sterlington Elementary School, 35% of them live east of Highway 165. Of the students who live east of Highway 165, 40% do NOT ride the bus. How many students who live east of Highway 165 ride the bus to school?

A. 75

B. 98

C. 147

D. 245

E. 273

30. What is the value of b in the solution of the system of equations below?
$$8a+4b=-8$$
$$-2a+b=10$$

F. -3

G. 0

H. 3

J. 4

K. 6

31. The graph below shows a relation of ordered pairs plotted in the standard (x,y) coordinate plane. Which of the following sets represents the domain of the relation shown?

A. $\{1, 3, 4\}$

B. $\{1, 2, 3, 4\}$

C. $\{1, 2, 3, 5\}$

D. $\{1, 2, 3, 6\}$

E. $\{1, 2, 3, 4, 5, 6\}$

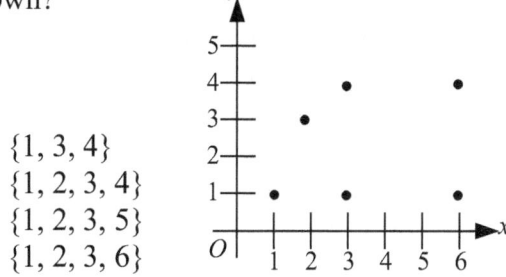

32. The figure below is made of 4 congruent line segments which intersect at right angles to form 12 congruent pieces. The sum of the lengths of the 12 congruent pieces is 48 inches. What is the area of the shaded region?

F. 4 in^2

G. 8 in^2

H. 16 in^2

J. 24 in^2

K. 36 in^2

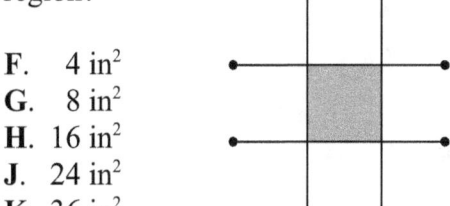

33. A school club is renting a charter bus for a field trip. The cost of the bus is $300 and will be shared equally among the students attending the trip. The current cost per student will decrease by $1 if 10 more students sign up to attend. How many students are currently signed up to attend?

A. 10

B. 15

C. 30

D. 50

E. 60

In the figure shown below, trapezoid *TRAP* is formed by Δ*TRP* and Δ*RAP*. The lengths are given in meters.

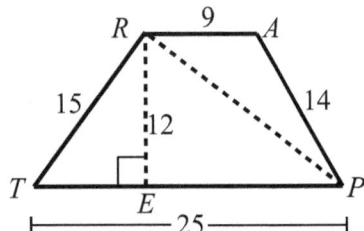

34. What is the area of Δ*TRP*, in square meters?
   F. 54
   G. 90
   H. 96
   J. 150
   K. 300

35. Which of the following ratios is equal to cos∠*EPR* ?

   A. $\dfrac{PR}{RT}$

   B. $\dfrac{PR}{PT}$

   C. $\dfrac{ER}{PR}$

   D. $\dfrac{ER}{EP}$

   E. $\dfrac{EP}{PR}$

36. Suppose *TRAP* is placed in the standard (*x,y*) coordinate plane such that *T* is at (0,0), *P* is at (25,0), and *R* and *A* are in the first quadrant. What is the *x*-coordinate of *A*?
   F. 9
   G. 16
   H. 18
   J. 21
   K. 25

37. In the figure shown below, Δ*ABC* ~ Δ*XYZ*, sides $\overline{AB}$ and $\overline{BC}$ are each 8.5 inches long, side $\overline{YZ}$ is 11.1 inches long, and the measure of ∠*A* is 71°. What is the measure of ∠*Y*?

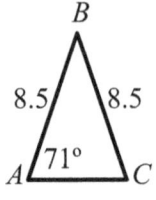
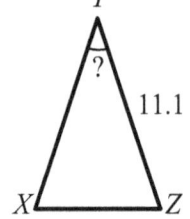

   A. 19°
   B. 38°
   C. 54.5°
   D. 60°
   E. 71°

38. Point *P* lies at (−3, 5) and point *Q* lies at (5, 8) in the standard (*x,y*) coordinate plane below. What is the length, in coordinate units, of $\overline{PQ}$?

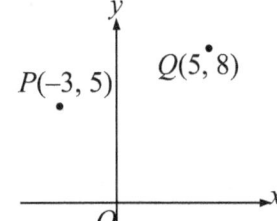

   F. $\sqrt{11}$
   G. $\sqrt{73}$
   H. $\sqrt{173}$
   J. 8
   K. 13

39. The figure below shows two concentric circles. The radius of the larger circle is 8 meters, and the radius of the smaller circle is 4 meters. What is the area, in square meters, of the shaded region between the two circles?

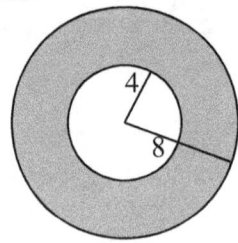

   A. 6π
   B. 12π
   C. 16π
   D. 48π
   E. 64π

40. The function $f(x) = .5\cos(2x)$ is graphed below for $0 \le x \le 2\pi$. What is the period of the function?

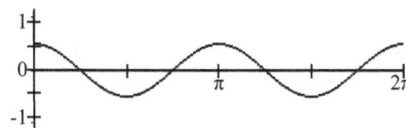

   F. $\dfrac{\pi}{4}$

   G. $\dfrac{\pi}{2}$

   H. π

   J. 2π

   K. 4π

41. If $f = \dfrac{2x+a}{2+b}$, what is x in terms of a, b, and f?

   A. $x = 2f + b - a$

   B. $x = bf + f - a$

   C. $x = \dfrac{2bf - a}{4}$

   D. $x = \dfrac{bf - a}{2}$

   E. $x = \dfrac{2f + bf - a}{2}$

42. For all $x \ne 3$, which of the following expressions is equal to $\dfrac{x^2 - 4x + 3}{x - 3} + 2x + 1$?

   F. $3x$

   G. $2x + 2$

   H. $2x^2 - x - 1$

   J. $\dfrac{3x}{x-3}$

   K. $\dfrac{x^2 - 2x + 4}{x - 3}$

43. For $i = \sqrt{-1}$, in standard form, $(2+i)^2 = ?$

   A. 3

   B. -3

   C. $3 + 4i$

   D. $5 + 4i$

   E. $4 + 4i + i^2$

> Use the following information to answer questions 44 – 46.

The points graphed below in the standard (x,y) coordinate plane below show the positions of 5 beehives located in a particular city. The positions are shown relative to the origin, which represents the intersection of large streets, and each coordinate unit represents 1 city block, which is equal to 150 meters.

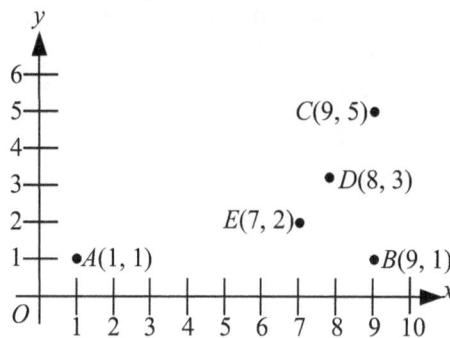

44. A group of college students studying the bees want to plant flowers at a central location and monitor the bees that visit. Since hives B, C, D, and E are clustered together, the students decide to find the average of the x-coordinates and the average of the y-coordinates of those four points to determine the best location to plant their flowers. What are the coordinates of this position?

   F. $(6.80, 2.40)$

   G. $(7.50, 2.50)$

   H. $(7.75, 3.00)$

   J. $(8.25, 2.50)$

   K. $(8.25, 2.75)$

45. What is the tangent of the angle formed by $\overrightarrow{AC}$ and $\overrightarrow{AB}$?

   A. $\dfrac{1}{2}$

   B. $\dfrac{\sqrt{5}}{2}$

   C. $\dfrac{\sqrt{5}}{5}$

   D. $\sqrt{5}$

   E. 2

46. Which of the following is closest to the number of meters between hive C and hive E?

- **F.** 13
- **G.** 130
- **H.** 212
- **J.** 541
- **K.** 1,950

_____

47. Which of the following expressions is the greatest monomial factor of $60\,a^5 b + 36\,a^3 b^2$?

- **A.** $12\,a^3 b$
- **B.** $12\,a^3 b^2$
- **C.** $12\,a^5 b^2$
- **D.** $90\,a^5 b^2$
- **E.** $90\,a^8 b^3$

48. In a jar of marbles, $\dfrac{1}{8}$ of the marbles are red, $\dfrac{1}{3}$ of the marbles are blue, $\dfrac{1}{4}$ of the marbles are yellow, and the remaining 21 marbles are green. How many red marbles are in the jar?

- **F.** 3
- **G.** 7
- **H.** 9
- **J.** 16
- **K.** 48

49. The volume of a square prism is $V = s^2 h$ where s is the side length of each side of the square base and h is the height of the prism perpendicular to the square base. Prism X and Prism Y are both square prisms. The side length of the square base on Prism X is 2 times the side length of the square base on Prism Y. The height of Prism X is half the height of Prism Y. Compared to Prism Y, the volume of Prism X is:

- **A.** the same.
- **B.** a quarter as great.
- **C.** half as great
- **D.** twice as great.
- **E.** four times as great.

50. Sydney and Benjamin begin to run from the same point on a track at the same time. Sydney runs at a constant rate of 75 seconds per lap, and Benjamin runs at a constant rate of 90 seconds per lap. How many seconds after beginning to run will Sydney have run exactly 1 more lap than Benjamin?

- **F.** 165
- **G.** 215
- **H.** 225
- **J.** 360
- **K.** 450

51. Which of the following number properties is illustrated in the statement below?
$$(2 + 5) + 9 = 2 + (5 + 9)$$

- **A.** Associative Property
- **B.** Commutative Property
- **C.** Distributive Property
- **D.** Inverse Property
- **E.** Transitive Property

52. In the standard (x,y) coordinate plane, for what value(s) of x, if any, is there NO value of y such that (x,y) is on the graph of
$$y = \frac{x^2 - 4}{(x+1)(x-1)(x-4)}?$$

- **F.** $-4, -1$, and 1 only
- **G.** $-1, 1$, and 4 only
- **H.** -2 and 2 only
- **J.** -1 and 1 only
- **K.** There are no such values of x.

53. If x is a positive even integer and y is a positive odd integer, then $[(+2)(-2)]^{xy}$ is:

- **A.** negative and odd
- **B.** negative and even
- **C.** zero
- **D.** positive and odd
- **E.** positive and even

54. Which of the following lists of numbers could be the side lengths of a triangle?

- **F.** 2 mm, 4 mm, 6 mm
- **G.** 3 in, 4 in, 7 in
- **H.** 2 ft, 9 ft, 12 ft
- **J.** 5 m, 10 m, 15 m
- **K.** 7 yd, 9 yd, 13 yd

55. The graph below shows an ellipse in the standard (x,y) coordinate plane. The center of the ellipse is $(0,0)$, and points $(0, 5)$, $(0, -5)$, $(3, 0)$, $(-3, 0)$, $A(a, 3)$, and $B(b, 3)$ lie on the ellipse. What is the distance, in coordinate units, from A to B?

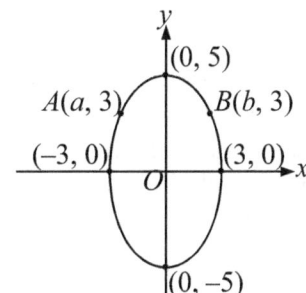

A. 2.4
B. 3.6
C. 4
D. 4.8
E. 6

56. Circle O is drawn in the standard (x,y) coordinate plane. 2 rays, $\overrightarrow{AB}$ and $\overrightarrow{AC}$ meet at A and are tangent to the circle at points B and C, respectively. Segment $\overline{BC}$ is drawn to create $\triangle ABC$. If $\angle A$ measures $40°$, what is the measure of $\angle ABC$?

F. 40°
G. 50°
H. 60°
J. 70°
K. Cannot be determined from the given information

57. In the standard (x,y) coordinate plane below, lines l, m, and n all have an x intercept of 3. the slope of line l is -1, the slope of line n is $-\dfrac{1}{4}$, and the slope of line m is the average of the slopes of lines l and line n. What is the y-intercept of line m?

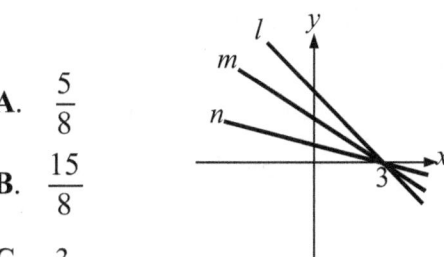

A. $\dfrac{5}{8}$

B. $\dfrac{15}{8}$

C. 3

D. $\dfrac{7}{2}$

E. 4

58. If n and x are positive rational numbers such that $n^{2x}=5$, then $n^{6x}=?$

F. 5
G. 15
H. 25
J. 30
K. 125

59. As shown in the figure below, $\overline{AB}$ and $\overline{CD}$ intersect at O. What is the value of x given that $0° < x° < 180°$ and $y = 5x$?

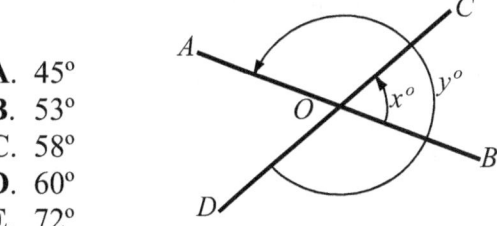

A. 45°
B. 53°
C. 58°
D. 60°
E. 72°

60. Assume x and y are distinct odd prime numbers and $z = 2x$. If $x \cdot y \cdot z$ is used as the common denominator when adding the fractions $\dfrac{1}{x}$, $\dfrac{1}{y}$, and $\dfrac{1}{z}$, then in order for the sum to be in lowest terms, its numerator and denominator must be reduced by a factor of which of the following?

F. 2
G. x
H. y
J. z
K. xy

Sample Test 2 Answers and Explanations

1. **D.** 54   If $l = 18$, then $w = \dfrac{1}{2}(18) = 9$. That means $P = 2l + 2w = 2(18) + 2(9) = 54$

2. **H.** 8 $|3 - 11| = |-8| = 8$

3. **A.** $2p - 9q$ Distribute the negative and combine like terms. $(8 - 6)p + (-5 - 4)q = 2p - 9q$

4. **H.** 7 $0.15(8) + 0.20(4) + 0.15x = 3.05$ $2 + 0.15x = 3.05$ $0.15x = 1.05$ $x = 7$

5. **A.** $0.15s + 110$ 15% of $s = 1.15s$ then add $110 to get $0.15s + 110$

6. **F.** $0.\overline{5}$ $0.\overline{5}$ means 5's repeat forever. All the other answers stop.

7. **E.** 22 in.   The difference is the end minus the beginning. $16 - (-6) = 22$

8. **H.** 3 $f(x) = 3x^2 + 2x - 5$ Plug in -2. $3(-2)^2 + 2(-2) - 5 = 12 - 4 - 5 = 3$

9. **E.** 74   Find the bottom angles of the triangle first. Then subtract from 180.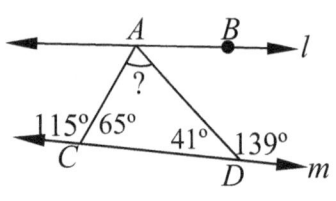
    $65 + 41 + ? = 180$
    $? = 74°$

10. **H.** 8   The median is the number in the middle when the numbers are in order. 5, 6, 6, (8,) 9, 10, 12

11. **C.** 38,000   2021 is 3 years after 2018, so $t = 3$. $P(3) = 25{,}000(1.15)^3 \approx 38000$

12. **J.** $2x^2 + 7x - 15$ $A = lw = (2x - 3)(x + 5)$ FOIL to get $2x^2 + 10x - 3x - 15$ and combine like terms.

13. **B.** II I III   Graph C is the only one going down, so it must be phrase III, the ball falling. Graph A is a straight line, so it must be phrase II, the constant rate. That leaves Graph B, which is curving upwards, to be phrase I, the accelerating train.

14. **G.** $\begin{bmatrix} 2 & 3 & | & 4 \\ 5 & -1 & | & 23 \end{bmatrix}$ An augmented matrix simply uses the numbers from the equations in standard form. The vertical line represents the equal signs. These equations are already in standard form, so don't move any numbers or change any signs.

15. **D.** 93   First find his current mean. $(80 + 75 + 90 + 87)/4 = 83$ He wants his mean to be two points higher, so that's 85. $(80 + 75 + 90 + 87 + x)/5 = 85$ and $x = 93$

16. **K.** $\overline{EF}$ All of the segments have positive slopes, so the one with the least value is the flattest one. There's no need to calculate anything.

17. **B.** $6,500   The total amount paid was $1500 + 250(60) = 18000$. To find how much more he paid, subtract the purchase price. $18000 - 11500 = 6500$

18. **F.** $\dfrac{6}{25}$ The probability is the ACME ball valves divided by the total. $\dfrac{12}{50} = \dfrac{6}{25}$

19. **C.** $\dfrac{3}{4}$ Use the Pythagorean Theorem to find the missing side. $20^2 + y^2 = 25^2$ so $y = 15$. Pitch is the same thing as slope. $\dfrac{\text{rise}}{\text{run}} = \dfrac{15}{20} = \dfrac{3}{4}$

20. **J.** 5   The y is greatest when the x is least. The least value x can have is 3. Plug in $3 + y \le 8$ and solve. $y \le 5$ so the greatest value for y is 5.

21. **E.** $\dfrac{7}{8} \cdot \dfrac{4}{5}$ If the area of the whole rectangle is 1, then assume the length and width are both 1. The length and width of the shaded area would then be $\dfrac{7}{8}$ and $\dfrac{4}{5}$.

22. **J.** $1,296   20% of 1500 is written as $0.20 \cdot 1500 = 300$, so the discounted price is $1500 - 300 = 1200$. The sales tax is

240

$0.08 \cdot 1200 = 96$, and the final price is $1200 + 96 = 1296$.

23. C. 18 $AD = AC + BD - BC$ because the little BC section was counted twice. Therefore $AD = 10 + 10 - 2 = 18$.

24. H. 125° $180 - 55 = 125$

25. C. 5.0   4 miles south and 3 miles west make a right angle. Use the Pythagorean Theorem to find the hypotenuse. $4^2 + 3^2 = 5^2$

26. G. $y = -\frac{1}{3}x - 1$ A slope of $-\frac{1}{3}$ means we can eliminate all but **G** and **H**. Plug in $(x, y) = (3, -2)$ and only **G** works.

27. A. $\frac{5}{3}$ $n \div \frac{1}{n} = n \cdot n = n^2 = \frac{25}{9}$ and $n = \sqrt{\frac{25}{9}}$

28. H. $\cos^{-1}\left(\frac{5}{13}\right)$ $\cos\theta = \frac{\text{adjacent}}{\text{hypotenuse}} = \frac{5}{13}$

29. C. 147   35% of 700 is $0.35 \cdot 700 = 245$ east of the highway. If 40% do not ride the bus, 60% ride the bus. $0.60 \cdot 245 = 147$

30. J. 4 $\begin{array}{l} 8a + 4b = -8 \\ 4(-2a + b = 10) \end{array}$ becomes

$\begin{array}{l} 8a + 4b = -8 \\ -8a + 4b = 40 \end{array}$ so $8b = 32$ and $b = 4$.

31. D. {1, 2, 3, 6}   The domain is the x values. There are points above 1, 2, 3, and 6.

32. H. 16 in²   If the sum of the lengths of the 12 pieces is 48, then each piece is 4. The area of the shaded square is $A = 4 \times 4 = 16$.

33. D. 50 $\frac{300}{x} - 1 = \frac{300}{x+10}$ $\frac{300}{x} - \frac{x}{x} = \frac{300}{x+10}$

$\frac{300-x}{x} = \frac{300}{x+10}$ cross multiply

$(300-x)(x+10) = 300x$

$3000 + 300x - 10x - x^2 = 300x$

$3000 - 10x - x^2 = 0$ so $x = -60$ or $x = 50$.

The answer can't be negative, so $x = 50$.

34. J. 150 $A = \frac{1}{2}bh = \frac{1}{2}25 \times 12 = 150$

35. E. $\frac{EP}{PR}$ $\cos\theta = \frac{\text{adjacent}}{\text{hypotenuse}} = \frac{EP}{PR}$

36. H. 18   First find TE. $TE^2 + 12^2 = 15^2$ $TE = 9$ and $RA = 9$, so the x coordinate of A is 18.

37. B. 38° $\triangle ABC$ is isosceles, so $\triangle XYZ$ is also isosceles. $\angle A$ is 71°, so $\angle C$, $\angle X$, and $\angle Z$ are also 71°. $71 + \angle Y + 71 = 180$ so $\angle Y = 38°$.

38. G. $\sqrt{73}$ The length in the x-direction is $5 - (-3) = 8$, and the length in the y-direction is $8 - 5 = 3$, so the diagonal length is $8^2 + 3^2 = d^2$ $d = \sqrt{73}$

39. D. 48π $A = \pi 8^2 - \pi 4^2 = 64\pi - 16\pi = 48\pi$

40. H. π A period is one full repeat of the graph. The section from 0 to π can be repeated to get the remainder of the graph, so it is one period.

41. E. $x = \frac{2f + bf - a}{2}$ Cross multiply

$\frac{f}{1} = \frac{2x+a}{2+b} \rightarrow 2f + bf = 2x + a \rightarrow$

$2f + bf - a = 2x \rightarrow \frac{2f + bf - a}{2} = x$

42. F. $3x$ $\frac{x^2 - 4x + 3}{x-3} + \frac{2x(x-3)}{x-3} + \frac{1(x-3)}{x-3}$

$\frac{x^2 - 4x + 3}{x-3} + \frac{2x^2 - 6x}{x-3} + \frac{x-3}{x-3} = \frac{3x^2 - 9x}{x-3}$

$= \frac{3x(x-3)}{x-3} = 3x$

43. C. $3 + 4i$ $(2+i)^2 = 4 + 2i + 2i + i^2 = 4 + 4i + (-1) = 3 + 4i$

44. K. (8.25, 2.75) $x = \frac{7+8+9+9}{4} = 8.25$ and

$y = \frac{1+2+3+5}{4} = 2.75$

45. A. $\frac{1}{2}$ $\tan\theta = \frac{\text{opposite}}{\text{adjacent}} = \frac{BC}{AB} = \frac{4}{8} = \frac{1}{2}$

46. J. 541   We need the distance between (9, 5) and (7, 2). The x-distance is $(9 - 7) = 2$ city blocks, and one city block equals 150 meters, so the x-distance is 300 meters. The y-distance is $(5 - 3) = 3$ city blocks, so the y-distance is 450 meters.

$d^2 = 300^2 + 450^2 = 292500$ and $d = \sqrt{292500} \approx 541$

47. A. $12a^3b$ The greatest common factor is the biggest number or variable that can be evenly divided into each term. 12 goes into 60 and 36 evenly, a^3 goes into a^5 and a^3 evenly, and b goes into b and b^2 evenly.

48. H. 9   First find the fraction of marbles that are green. Subtract from 1 whole jar.

$1-\dfrac{1}{8}-\dfrac{1}{3}-\dfrac{1}{4}=\dfrac{7}{24}$ If $\dfrac{7}{24}$ of the marbles are green, and there are 21 green marbles, we can find the total number of marbles by solving $\dfrac{7}{24}x=21$. There are 72 marbles.

$\dfrac{1}{8}\cdot72=9$ red marbles.

49. D. twice as great.   Make up some easy numbers. Every side on Prism Y is 1. Therefore, its volume is 1. The sides on Prism X are twice as long, so $s = 2$. The height is half as tall, so $h = 0.5$.

$V=2^2\cdot0.5=2$ The volume of Prism X is 2, which is twice as great as 1.

50. K. 450   Notice that the rates are in seconds per lap, not laps per second. If x is seconds run, Sydney's laps = $x/75$ and Benjamin's laps = $x/90$. We want Sydney to equal Benjamin + 1. $\dfrac{x}{75}=\dfrac{x}{90}+1$ Solve. $x=450$

51. A. Associative Property   The order of the numbers didn't change, only the ones that were grouped (associated) together.

52. G. −1, 1, and 4 only   These are the three values that make the bottom of the fraction equal zero. Anything over zero is undefined.

53. E. positive and even   If x is a positive even integer and y is a positive odd integer, then xy is a positive even integer. −4 raised to any positive integer power has to be even. A negative raised to an even power is positive.

54. K. 7 yd, 9 yd, 13 yd   The sum of the two shorter sides of a triangle must be larger than the longest side, otherwise the short sides won't meet in the middle. If the sum is equal to the long side, the triangle can't bend, and the result is just a line segment. $7+9>13$

55. B. 3.6   The equation of the ellipse is

$\dfrac{x^2}{3^2}+\dfrac{y^2}{5^2}=1$. For a and b, the y value is 3.

Plug in to find the two possible values for x.

$\dfrac{x^2}{3^2}+\dfrac{3^2}{5^2}=1 \;\rightarrow\; \dfrac{x^2}{9}=\dfrac{9}{25} \;\rightarrow\; x^2=\dfrac{81}{25} \;\rightarrow\;$

$x=\pm\dfrac{9}{5}$ The distance between the two values is $\dfrac{18}{5}$ or 3.6.

56. J. 70° $\triangle ABC$ is an isosceles triangle. If $\angle A$ measures 40°,
$40+2x=180$ and $x = 70°$.

57. B. $\dfrac{15}{8}$ The slope of line m is

$\left(-\dfrac{1}{4}+(-1)\right)\div2=-\dfrac{5}{8}$ and it passes

through the point (3, 0). The equation of the line is $y-0=-\dfrac{5}{8}(x-3)$, which simplifies

to $y=-\dfrac{5}{8}x+\dfrac{15}{8}$. The y-intercept is $\dfrac{15}{8}$.

58. K. 125 $n^{2x}=5 \;\rightarrow\; \left(n^{2x}\right)^3=5^3 \;\rightarrow\;$
$n^{6x}=125$

59. D. 60°   The unlabeled angle on the left is also equal to x, so $x + y = 360$ and $y = 5x$.
$x+5x=360$, so $x = 60°$.

60. G. x $\dfrac{1}{x}+\dfrac{1}{y}+\dfrac{1}{z}=\dfrac{yz+xz+xy}{xyz}$ If we

substitute $z = 2x$, we get $\dfrac{2xy+2x^2+xy}{2x^2y}$.

Factor $\dfrac{x(2y+2x+y)}{2x^2y}$ and we can cancel

an x. The LCD of x, y, and $2x$ is $2xy$. We don't need the other x.

Sample Test 3

1. A game has a regular price of $49.95 before taxes. It goes on sale at 20% below the regular price. Before taxes are added, what is the sale price of the game?
   A. $ 9.99
   B. $24.98
   C. $29.95
   D. $39.96
   E. $49.75

2. Two dials are shown below. When the arrow on each dial is spun, it is equally likely to point at any of the numbered sectors on its dial after it has stopped spinning. After the arrows are next spun, the numbers in the sectors at which the arrows point after they stop spinning will be added together. Which of the following values is NOT a possible sum of those two numbers?

    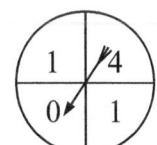

   F. 1
   G. 2
   H. 3
   J. 4
   K. 5

3. To produce football helmets, it costs Hard Head Equipment Company $2,100 for overhead, plus $6 per helmet produced. What is the maximum number of football helmets that can be produced by the company for $18,000?
   A. 1,450
   B. 2,000
   C. 2,118
   D. 2,500
   E. 2,650

4. If $\frac{2}{5}x + 12 = 25$, then x = ?

   F. $-\dfrac{65}{2}$

   G. $\dfrac{5}{2}$

   H. $\dfrac{26}{5}$

   J. $\dfrac{65}{2}$

   K. $\dfrac{185}{2}$

5. In the figure below, C lies on $\overline{BD}$, $\angle CAD$ measures 34°, and $\angle ADC$ measures 122°. What is the measure of $\angle ACB$?

   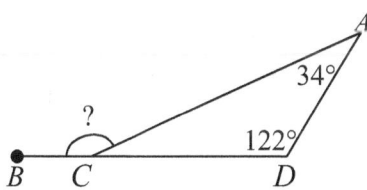

   A. 122°
   B. 128°
   C. 134°
   D. 146°
   E. 156°

6. In the figure below, a circle with a radius of 14 mm circumscribes a regular hexagon. What is the perimeter, in mm, of the hexagon?
   F. 42
   G. $42\sqrt{3}$
   H. 84
   J. $84\sqrt{2}$
   K. $84\sqrt{3}$

7. What is the length, in meters, of the hypotenuse of a right triangle with legs that are 7 meters long and 3 meters long?

A. $2\sqrt{5}$

B. $2\sqrt{10}$

C. $\sqrt{58}$

D. 10

E. 21

8. A bucket contains exactly 22 bottle caps: 4 yellow, 6 red, and 12 green. What is the probability of randomly selecting 1 bottle cap that is NOT green?

F. $\dfrac{1}{22}$

G. $\dfrac{5}{11}$

H. $\dfrac{6}{11}$

J. $\dfrac{2}{3}$

K. $\dfrac{5}{6}$

9. For all x such that $x \neq 0$, which of the following expressions is equivalent to $\dfrac{16x^2+8x}{4x}$?

A. $6x$

B. $12x$

C. $4x+2$

D. $4x^2+2$

E. $16x^2+2$

10. Given that $3x + 4 = 2$ and $4y + 3 = 5$, what is $x + y$?

F. $-\dfrac{1}{3}$

G. $-\dfrac{1}{6}$

H. $\dfrac{1}{6}$

J. $\dfrac{5}{6}$

K. 1

11. What is the sum of three consecutive even integers whose mean is 36?

A. 72

B. 102

C. 108

D. 114

E. 180

12. Given $a = 5$, $c = 6$, and $t = -8$, $(a+c-t)(c+t)=?$

F. –38

G. –6

H. 6

J. 14

K. 42

13. An endurance race has 5 checkpoints as shown below. The race designers have posted the following distances: 11.4 miles between checkpoints 1 and 3, 6.3 miles between checkpoints 2 and 3, and 11.2 miles between checkpoints 2 and 5. Which of the following values is closest to the distance, in miles, between checkpoints 1 and 5?

A. 16.3

B. 17.5

C. 17.7

D. 22.6

E. 28.9

14. The graph below shows the water level in a rain gauge over a period of 7 hours. One of the following values shows the number of hours the water level remained constant. Which one?

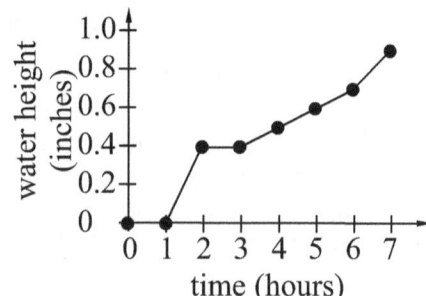

F. 2

G. 3

H. 4

J. 4.5

K. 7

244

15. If a student makes an A on a particular test, the probability that the student makes an A on the following test is 0.8. If a student does not make an A on a particular test, the probability that the student will make an A on the following test is 0.3. Given that a student made an A on the last test, what is the probability that the student will NOT make an A on the following test?

A. 0.2
B. 0.3
C. 0.5
D. 0.7
E. 0.8

16. In the figure below, line k is parallel to line l, and line m is perpendicular to line n. The acute angle formed by lines k and m measures 52°. What is the value of x?

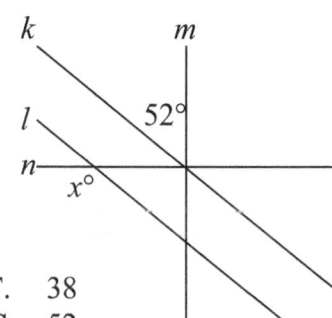

F. 38
G. 52
H. 128
J. 132
K. 142

17. One of the following equations represents the line graphed in the standard (x,y) coordinate plane below. Which one?

A. $y=-\dfrac{3}{4}x+3$

B. $y=-\dfrac{4}{3}x+3$

C. $y=-\dfrac{3}{4}x+4$

D. $y=\dfrac{3}{4}x+3$

E. $y=\dfrac{4}{3}x+4$

18. What is the value of the expression
$$\dfrac{|-6-2|^2+(-2)^3}{12\div3\times2}?$$

F. −7
G. 0
H. 7
J. 8
K. 28

19. Kyle invested $1,500 in a special savings account. The balance of this account will double every 7 years. Assuming Kyle makes no other deposits or withdrawals, what will be the balance of Kyle's investment at the end of 42 years?

A. $ 21,000
B. $ 42,000
C. $ 63,000
D. $ 96,000
E. $1,920,000

20. Given that the function f defined as $f(x)=3-4x$ has domain {−2, 0, 2}, what is the range of f?

F. {−2, 0, 2}
G. {−5, −1, 11}
H. {−5, 3, 11}
J. {−8, −1, 8}
K. {−8, 3, 11}

21. A jar contains 14 solid-colored buttons of the same size: 2 red, 3 brown, 4 blue, and 5 black. Which of the following expressions gives the probability of drawing, at random and without replacement, a black button on the first draw, a red button on the second draw, and a black button on the third draw?

A. $\left(\dfrac{5}{14}\right)\left(\dfrac{2}{14}\right)\left(\dfrac{4}{14}\right)$

B. $\left(\dfrac{5}{14}\right)\left(\dfrac{2}{14}\right)\left(\dfrac{4}{14}\right)$

C. $\left(\dfrac{5}{14}\right)\left(\dfrac{2}{13}\right)\left(\dfrac{5}{12}\right)$

D. $\left(\dfrac{5}{14}\right)\left(\dfrac{4}{13}\right)\left(\dfrac{3}{12}\right)$

E. $\left(\dfrac{5}{14}\right)\left(\dfrac{2}{13}\right)\left(\dfrac{4}{12}\right)$

245

22. A statistics teacher recorded the grades each of his students earned on their first statistics test. The test scores of each of the students are represented in the stem-and-leaf plot shown below.

Stem	Leaf
6	5 8 8
7	2 3 9
8	4 5 6 8 8 9
9	0 1 1 6

What is the probability that a student chosen at random will have earned *at least* an 85 on the first statistics test?

F. $\dfrac{5}{20}$

G. $\dfrac{9}{20}$

H. $\dfrac{5}{16}$

J. $\dfrac{8}{16}$

K. $\dfrac{9}{16}$

23. Two bicycle tires have radii of 9 in and 13 in, respectively. How many inches longer is the circumference of the larger tire than that of the smaller tire?

A. 4
B. 4π
C. 8π
D. 22π
E. 88π

24. In the standard (x,y) coordinate plane, what is the slope of the line given by the equation $3x = 4y + 24$?

F. $-\dfrac{3}{4}$

G. $\dfrac{3}{4}$

H. $\dfrac{4}{3}$

J. 6

K. 8

25. To the nearest foot, what is the height of a rectangular prism with base length 18 ft, base width of 1.5 ft, and volume of 85 ft³?

A. 3
B. 7
C. 9
D. 28
E. 46

26. In the figure shown below, a ladder 12 feet long leans against a vertical building making a 58° angle with the ground. Which of the following expressions represents the distance, in feet, along the building between the ground and the top of the ladder?

F. $\dfrac{12}{2}$

G. $\dfrac{12\sqrt{3}}{2}$

H. 12 sin 58°
J. 12 cos 58°
K. 12 tan 58°

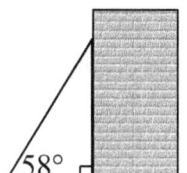

27. The isotope vanadium-48 has a half-life of 16 days, which means that the amount of vanadium-48 remaining after t days is $N\left(\dfrac{1}{2}\right)^{\frac{t}{16}}$, where N is the number of grams of vanadium-48 at $t=0$. How many grams of vanadium-48 will remain after 32 days if there were 64 grams of vanadium-48 at $t=0$?

A. 0
B. 16
C. 32
D. 128
E. 256

28. Which of the following expressions is equal to $\sqrt[4]{625x^{12}}$?

F. $5x^3$
G. $5x^8$
H. $25x^3$
J. $25x^8$
K. $125x^6$

29. Values of the functions f and g are shown below in tables. What is the value of $f(g(0))$?

x	$f(x)$
3	8
4	2
5	4
6	1

x	$g(x)$
–2	3
0	5
2	4
5	6

A. 0
B. 1
C. 2
D. 4
E. 8

30. The figure below shows a rectangle measuring 6 mm by 15 mm which has had a square removed from each corner. The squares each have sides of length x mm. In terms of x, what is the area, in mm^2, of the rectangle?

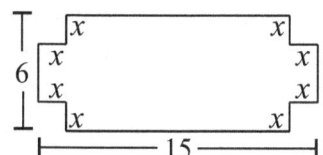

F. $42-4x^2$
G. $90-4x^2$
H. $90+4x^2$
J. $90-4x$
K. $90-42x+4x^2$

31. In the standard (x,y) coordinate plane below, square $ABCD$ will be reflected over the x-axis and then translated 8 units right. What will be the coordinates of the final image of A resulting from both transformations?

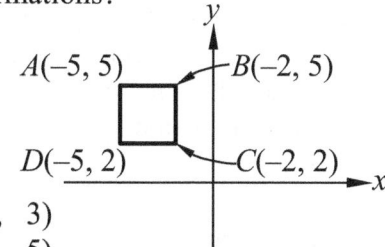

A. $(-5,\ 3)$
B. $(-3,\ 5)$
C. $(-3,\ 8)$
D. $(\ 3,\ 8)$
E. $(\ 3,-5)$

Use the following information to answer questions 32 – 34.

In the standard (x,y) coordinate plane below, $\triangle AOB$ is formed by $\overleftrightarrow{AB}$, the x-axis, and the y-axis.

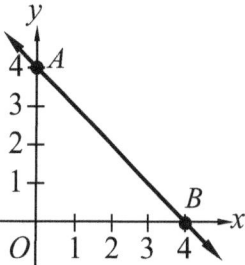

32. What is the length of $\overline{AB}$ in coordinate units?
F. $2\sqrt{2}$
G. $4\sqrt{2}$
H. $4\sqrt{3}$
J. 4
K. 8

33. What is the area of $\triangle AOB$ in square coordinate units?
A. 4
B. $4\sqrt{2}$
C. 8
D. $8\sqrt{2}$
E. 16

34. Which of the following is an equation of $\overleftrightarrow{AB}$?
F. $y=\ -x+4$
G. $y=\ x-4$
H. $y=\ x+4$
J. $y=-4x-4$
K. $y=\ 4x+4$

35. What is the solution set of the equation $x^4+7x^2-144=0$?
A. $\{-16,\ 9\}$
B. $\{-16,-3,\ 3\}$
C. $\{\ -9,-4,\ 4\}$
D. $\{\ -4,\ 4,-3i,\ 3i\}$
E. $\{\ -3,\ 3,-4i,\ 4i\}$

36. Which of the following arranges the numbers $\frac{7}{4}$, $1.\overline{7}$, $1.\overline{75}$, and 1.8 into ascending order?

F. $\frac{7}{4} < 1.\overline{7} < 1.\overline{75} < 1.8$

G. $\frac{7}{4} < 1.\overline{75} < 1.\overline{7} < 1.8$

H. $1.\overline{7} < 1.\overline{75} < \frac{7}{4} < 1.8$

J. $1.\overline{75} < 1.\overline{7} < 1.8 < \frac{7}{4}$

K. $1.\overline{75} < 1.\overline{7} < \frac{7}{4} < 1.8$

37. Maura, Riley, and Shelly are standing in a meadow such that Maura is 60 feet due North of Riley and Shelly is 80 feet due west of Riley. If their positions are drawn as vertices of a triangle, which of the following expressions represents the degree measure of the angle of the triangle at the vertex where Shelly is standing?

A. $\sin^{-1}\left(\frac{60}{80}\right)$

B. $\sin^{-1}\left(\frac{80}{60}\right)$

C. $\cos^{-1}\left(\frac{60}{80}\right)$

D. $\tan^{-1}\left(\frac{60}{80}\right)$

E. $\tan^{-1}\left(\frac{80}{60}\right)$

38. Which of the following operations will produce the largest result when substituted for the blank in the expression

$15\underline{\quad}\left(\frac{-1}{30}\right)$?

F. Plus
G. Minus
H. Multiplied by
J. Divided by
K. Averaged with

39. Amari's e-shop has been reviewed by 50 customers. The table below summarizes the number of stars given by the customers.

Rating (Stars)	Number of customers
5	26
4	11
3	2
2	3
1	8

Which of the following values is closest to the mean of the 50 customer ratings?

A. 1.3
B. 3.0
C. 3.6
D. 3.9
E. 4.8

40. A local consignment store has a sale every December. Each item is discounted by 5% for every month it has been at the store. For example, an item that arrived in January has been there for 11 months and would be discounted by 55%. If D represents the item's discount during the December sale, and M represents the month the item arrived, which of the following equations gives D in terms of M?

F. $D = 0.05M$
G. $D = 0.55M$
H. $D = 0.6 - M$
J. $D = 0.6 + 0.05M$
K. $D = 0.6 - 0.05M$

41. The equation $e = -0.01089\,m + 44.15$ models the highway gas mileage of a vehicle, e, in miles per gallon, based on the vehicle mass, m kilograms. According to this equation, what is the expected gas mileage of a vehicle weighing 2000 kilograms?

A. 22.37 mpg
B. 33.26 mpg
C. 41.97 mpg
D. 44.04 mpg
E. 44.15 mpg

42. A semicircular window is shown below. Which of the following values is closest to the area, in square inches, of the window?

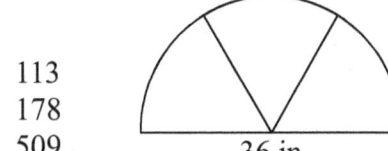

36 in

F. 113
G. 178
H. 509
J. 648
K. 1018

43. Which of the following is the equation of a circle in the standard (x,y) coordinate plane with the center located at $(-3, 2)$ and a radius of 6 coordinate units?

A. $(x+3)^2+(y-2)^2=6$
B. $(x-3)^2+(y+2)^2=6$
C. $(x-3)^2+(y-2)^2=\sqrt{6}$
D. $(x+3)^2+(y-2)^2=36$
E. $(x-3)^2+(y+2)^2=36$

44. River City Doggie Day Care is building a new outdoor puppy pen. They have 72 feet of fencing, and they plan to buy a 4-foot-wide gate. Which of the following options will give them a pen with the largest area if only the fencing and gate are used to enclose it?

F. A square with a side length of 18 feet.
G. A square with a side length of 19 feet.
H. A rectangle with a length of 19 feet and a width of 17 feet.
J. A rectangle with a length of 20 feet and a width of 18 feet.
K. A rectangle with a length of 22 feet and a width of 18 feet.

45. If $25^a=5$ and $2^{a+b}=8$, then $b=$?

A. $\dfrac{1}{2}$
B. $\dfrac{3}{2}$
C. $\dfrac{5}{2}$
D. 3
E. $\dfrac{7}{2}$

46. What is the smallest positive integer having exactly 5 different positive integer divisors?

F. 6
G. 8
H. 12
J. 16
K. 120

47. The difference $\dfrac{2}{5}-\dfrac{-1}{2}$ lies in which of the following intervals graphed on the real number line?

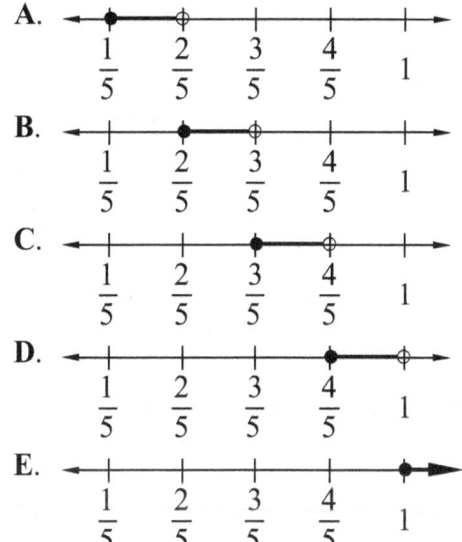

48. What is the common difference of an arithmetic sequence if the first term is -3 and the sixteenth term is 117?

F. 7
G. 7.5
H. 8
J. 9
K. 13

49. How many integers between, but not including, 30 and 40 have a prime factorization with exactly 3 factors that are NOT necessarily unique? (Note: 1 is NOT a prime number.)

A. 0
B. 1
C. 2
D. 3
E. 4

Use the following information to answer questions 50 – 52.

Big Easy Gators is offering swamp tours. The tables below give information about the swamp boats and the tours offered.

Swamp Boat Information	
Overall length	22 feet
Max capacity	8 passengers
Tour operator	Boudreaux
Overall length	19 feet
Max capacity	6 passengers
Tour operator	Bubba

Tour Information			
Tour	Cost	Duration	Operator
A	$100	30 min	Bubba
B	$125	45 min	Bubba
C	$150	90 min	Boudreaux

50. Josephine went on Tour B, and her trip covered a distance of 9 miles. Philip went on Tour C, and his trip covered a distance of 22.5 miles. Which of the following values is the difference, in miles per hour, of the average speeds of their boats during their tours?
F. 0
G. 3
H. 9
J. 12.4
K. 13.5

51. Big Easy Gators made $9,400 in 1 day by selling a total of 71 tickets for their three tours. They sold twice as many tickets for Tour A as for Tour B. How many tickets were sold for Tour C?
A. 10
B. 15
C. 20
D. 30
E. 41

52. Bubba swerves his boat to just miss an alligator's tail. At the moment the front of the boat passes the tail, the alligator is perpendicular to the boat's direction of travel, and the angle from Bubba to the alligator's nose is 48°. Which of the following expressions is closest to the length, l feet, from the alligator's nose to its tail?

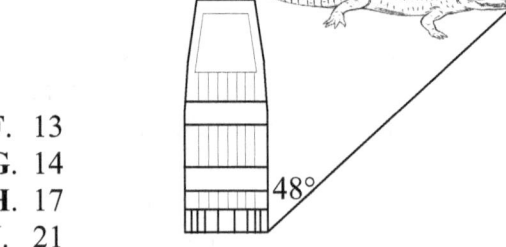

F. 13
G. 14
H. 17
J. 21
K. 24

53. The graphs of $y=x+1$, $y=-x-1$, and $x^2+y^2=4$ are shown in the standard (x,y) coordinate plane below. The shaded region is the solution to one of the following systems of inequalities. Which system is it?

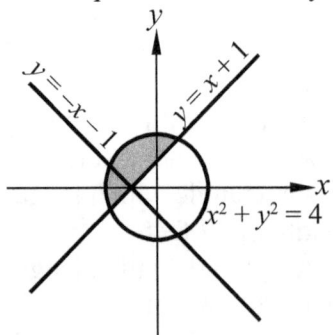

A. $\begin{array}{l} y \geq x+1 \\ x^2+y^2 \geq 4 \end{array}$

B. $\begin{array}{l} y \geq -x-1 \\ x^2+y^2 \geq 4 \end{array}$

C. $\begin{array}{l} y \leq x+1 \\ x^2+y^2 \leq 4 \end{array}$

D. $\begin{array}{l} y \geq x+1 \\ x^2+y^2 \leq 4 \end{array}$

E. $\begin{array}{l} y \geq -x-1 \\ x^2+y^2 \leq 4 \end{array}$

54. A city park is in the shape of a triangle which is shown below. A landscaper plans to plant a tree about every 6 meters along the perimeter of the park. Among the following, which expression best estimates the planned number of trees the landscaper intends to plant along the perimeter of the park?

(Note: The law of sines states that in every triangle, the three ratios of length of a side to the sine of the angle opposite that side are equal.)

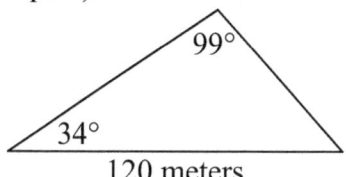

F. $\dfrac{120+\left(\dfrac{120\sin 34°}{\sin 99°}\right)+\left(\dfrac{120\sin 47°}{\sin 99°}\right)}{6}$

G. $\dfrac{120+\left(\dfrac{120\sin 99°}{\sin 34°}\right)+\left(\dfrac{120\sin 99°}{\sin 47°}\right)}{6}$

H. $120+\dfrac{120\sin 34°}{\sin 99°}+\dfrac{120\sin 47°}{\sin 99°}$

J. $\dfrac{\dfrac{1}{2}\left(\dfrac{120\sin 47°}{\sin 99°}\right)}{6}$

K. $\dfrac{\dfrac{1}{2}(120)}{6}$

55. The probabilities that each of two independent events will occur are given in the table below.

Event	Probability
A	0.30
B	0.25

What is the probability that both events will occur, that is, $P(A \text{ and } B)$?

A. 0.075
B. 0.25
C. 0.275
D. 0.30
E. 0.55

56. The function $f(x)$ is shown below with several points labeled. Another function, $g(x)$, is defined such that $g(x)=5-f(x)$. What is $g(-2)$?

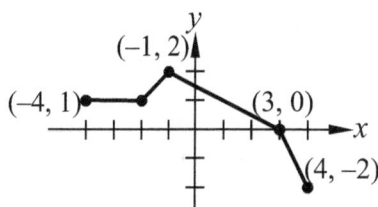

F. -4
G. -1
H. 1
J. 4
K. 6

57. The ratio of a to b is 4 to 1, and the ratio of b to c is 15 to 1. What is the value of $\dfrac{2a+6b}{3b+5c}$?

A. $\dfrac{28}{15}$

B. $\dfrac{21}{5}$

C. $\dfrac{29}{3}$

D. $\dfrac{33}{1}$

E. $\dfrac{127}{1}$

58. In the standard (x,y) coordinate plane, what is the y-intercept of the function $y=f(x)$ defined below?

$$f(x)=\begin{cases} x^2-8 & \text{for } x<-5 \\ 3x+4 & \text{for } -5\le x\le 6 \\ |2x-2| & \text{for } x>6 \end{cases}$$

F. -8
G. -5
H. -2
J. 4
K. 6

59. The frequency histogram below shows the sale price of 9 trading cards from an online auction site.

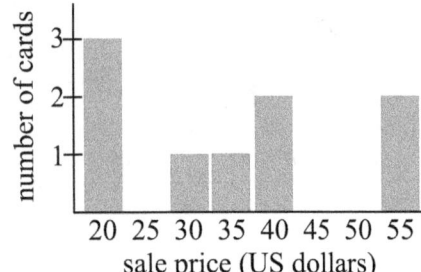

Using the data from the frequency histogram, what is the sum of the mean and the median of this distribution?

A. 65
B. 70
C. 75
D. 80
E. 85

60. What is the matrix product $\begin{bmatrix} 2 & 3 \\ 4 & 5 \end{bmatrix}\begin{bmatrix} a & b \\ c & d \end{bmatrix}$?

F. $\begin{bmatrix} 2a & 3b \\ 4c & 5d \end{bmatrix}$

G. $\begin{bmatrix} (2a+3b) \\ (4c+5d) \end{bmatrix}$

H. $\begin{bmatrix} (2a+4c) & (3b+5d) \end{bmatrix}$

J. $\begin{bmatrix} (2a+4b) & (3a+5b) \\ (2c+4d) & (3c+5d) \end{bmatrix}$

K. $\begin{bmatrix} (2a+3c) & (2b+3d) \\ (4a+5c) & (4b+5d) \end{bmatrix}$

Sample Test 3 Answers and Explanations

1. **D.** $39.96   20% of $49.95 is
$0.20 \times 49.95 = 9.99$ and
$49.95 - 9.99 = 39.96$

2. **J.** 4   It is impossible to get a sum of 4.
$1+0=1$, $1+1=2$, $2+1=3$, and $1+4=5$.

3. **E.** 2,650 $2,100+6x=18,000$
$6x=15900$ so $x=2650$

4. **J.** $\dfrac{65}{2}$ $\dfrac{2}{5}x+12=25$ Subtract the 12.

$\dfrac{2}{5}x=13$ Multiply by 5. $2x=65$ Divide by

2. $x=\dfrac{65}{2}$

5. **E.** 156°   Find the third angle of the
triangle. $180 - 34 - 122 = 24$
$180 - 24 = 156$

6. **H.** 84   The hexagon can be divided into six
congruent triangles. The central angles each
measure $360/6 = 60°$. This
means they are equilateral
triangles and each side
measures 14. The perimeter is
$14 \times 6 = 84$.

7. **C.** $\sqrt{58}$ Use the
Pythagorean Theorem. $7^2+3^2=c^2$
$49+9=c^2$ $58=c^2$ $\sqrt{58}=c$

8. **G.** $\dfrac{5}{11}$ There are 10 bottle caps that are

not green, so the probability is $\dfrac{10}{22}=\dfrac{5}{11}$

9. **C.** $4x+2$ Factor the expression.
$\dfrac{16x^2+8x}{4x}=\dfrac{8x(2x+1)}{4x}$ Then reduce.

$\dfrac{8x(2x+1)}{4x}=2(2x+1)=4x+2$

10. **G.** $-\dfrac{1}{6}$ First, solve for x. $3x+4=2$

$3x=-2$ $x=-\dfrac{2}{3}$ Then solve for y.

$4y+3=5$ $4y=2$ $y=\dfrac{1}{2}$ Add x and y.

$-\dfrac{2}{3}+\dfrac{1}{2}=-\dfrac{4}{6}+\dfrac{3}{6}=-\dfrac{1}{6}$

11. **C.** 108   Three consecutive even integers
can be represented as x, $x + 2$, and $x + 4$.
Add 2 each time to make them all even (or
odd) integers. $\dfrac{x+x+2+x+4}{3}=36$

$\dfrac{3x+6}{3}=36$ $3x+6=108$ (You may notice

that this is the sum. You're done!
Otherwise...) $3x=102$ $x=34$ $x+2=36$
and $x+4=38$ $34+36+38=108$

12. **F.** –38   Plug in the numbers and simplify.
$(5+6-(-8))(6+(-8))=(11+8)(6-8)$
$(19)(-2)=-38$

13. **A.** 16.3   The distance from 1 to 3 is 11.4
miles, and the distance from 2 to 3 is 6.3
miles. We can subtract to find the distance
from 1 to 2 is 5.1 miles. The distance from
2 to 5 is 11.2 miles. We don't need to know
the other individual distances. We can just
add 1 to 2 and 2 to 5 to get 16.3 miles.

14. **F.** 2   A graph is constant when it is
horizontal. This graph is constant from 0 to
1 and from 2 to 3. That's 2 hours.

15. **A.** 0.2   If a student made an A on the last
test, the probability of making an A on the
next test is 0.8. This means the probability
of not making an A is $1 - 0.8 = 0.2$

16. **K.** 142   If line m were
removed, the obtuse
angle would be
$52 + 90 = 142°$.
Angle x is congruent
to this angle.

253

17. A. $y=-\dfrac{3}{4}x+3$ The slope is negative because the line goes down from left to right. $m=\dfrac{rise}{run}=-\dfrac{3}{4}$. The y-intercept is 3.

18. H. 7 $\dfrac{|-6-2|^2+(-2)^3}{12\div3\times2}=\dfrac{|-8|^2+(-2)^3}{4\times2}$

$\dfrac{64+(-8)}{8}=\dfrac{56}{8}=7$

19. D. $96,000   Since the account doubles every 7 years, in 42 years it will double 6 times. Multiply $1,500 by 2 six times. $1,500(2)^6=96,000$

20. H. {–5, 3, 11}   Plug in each value of the domain. $3-4(-2)=11$, $3-4(0)=3$, and $3-4(2)=-5$

21. E. $\left(\dfrac{5}{14}\right)\left(\dfrac{2}{13}\right)\left(\dfrac{4}{12}\right)$ For the first draw, there are 5 black buttons out of 14 total $\left(\dfrac{5}{14}\right)$. For the second draw, there are 2 red buttons out of 13 total $\left(\dfrac{2}{13}\right)$. For the third draw, there are 4 black buttons out of 12 total $\left(\dfrac{4}{12}\right)$.

22. K. $\dfrac{9}{16}$ There are 16 scores listed, and 9 of them (96, 91, 91, 90, 89, 88, 88, 86, and 85) are at least 85. The probability is $\dfrac{9}{16}$.

23. C. 8π $C=2\pi r$. The circumference of the smaller tire is $C=2\pi9=18\pi$. The circumference of the larger tire is $C=2\pi13=26\pi$. $26\pi-18\pi=8\pi$.

24. G. $\dfrac{3}{4}$ Rearrange the equation to make it slope-intercept form. $3x-24=4y$ $y=\dfrac{3}{4}x-6$ The slope is $\dfrac{3}{4}$.

25. A. 3   Volume is length times width times height. Plug in and sole for h. $V=lwh$ $85=(18)(1.5)(h)$ $85=27h$ $h=3$

26. H. 12 sin 58°   The side along the building is the opposite, and the ladder is the hypotenuse. $\sin58°=\left(\dfrac{opposite}{hypotenuse}\right)=\dfrac{x}{12}$, so $x=\sin58°$.

27. B. 16 $N=64$ and $t=32$. $64\left(\dfrac{1}{2}\right)^{\frac{32}{16}}=64\left(\dfrac{1}{2}\right)^2=64\left(\dfrac{1}{4}\right)=16$

28. F. $5x^3$ A fourth root means four identical factors are needed to come out of the radical. $625=5\times5\times5\times5$, so $\sqrt[4]{625}=5$. $x^{12}=x^3\times x^3\times x^3\times x^3$, so $\sqrt[4]{x^{12}}=x^3$.

29. D. 4   Start from the inside and work out. Use the second chart to find $g(0)=5$, so $f(g(0))=f(5)$. Use the first chart to find $f(5)=4$.

30. G. $90-4x^2$ The area of the shape is equal to the area of the whole rectangle minus the four corners. The area of each missing corner is $x\times x=x^2$, and the area of the whole rectangle is $15\times6=90$. The final area is $90-4x^2$.

31. E. (3, –5)   Reflecting over the x-axis results in moving point A down to (–5, –5). Then, shifting right 8 units will move it to (3, –5).

32. G. $4\sqrt{2}$ Use the Pythagorean theorem. The two legs are each 4 units long. $4^2+4^2=c^2$ so $32=c^2$ and $c=\sqrt{32}=4\sqrt{2}$.

33. C. 8 $A=\dfrac{1}{2}bh$ and the two legs are each 4 units long. $A=\dfrac{1}{2}(4)(4)=8$

34. F. $y=-x+4$ The slope is negative because the line goes down from left to right. $m=\dfrac{rise}{run}=-\dfrac{4}{4}=-1$. The y-intercept is 4.

35. E. {–3, 3, –4i, 4i}   Factor the equation. $x^4+7x^2-144=(x^2-9)(x^2+16)$. The difference of squares factors normally $x^2-9=(x+3)(x-3)$. The sum of squares gives imaginary numbers $x^2+16=(x+4i)(x-4i)$. Finally, set each factor equal to zero and solve for x. $x+3=0$ so $x=-3$, etc.

36. G. $\dfrac{7}{4}<1.\overline{75}<1.\overline{7}<1.8$ Convert the fraction to a decimal. $\dfrac{7}{4}=1.75$ Pay

attention to which decimals repeat.
1.7500<1.7575<1.7777<1.8000

37. D. $\tan^{-1}\left(\dfrac{60}{80}\right)$ 60 is the opposite side, and 80 is the adjacent side. The hypotenuse is not given. $\tan x = \dfrac{opposite}{adjacent} = \dfrac{60}{80}$ so $x = \tan^{-1}\left(\dfrac{60}{80}\right)$

38. G. Minus   Multiplying and dividing will result in a negative answer. Averaging will give an answer between the two numbers. Adding a negative will give a number a little smaller than 15. Minus a negative will add 15 and $\dfrac{1}{30}$, giving the largest result.

39. D. 3.9 $\dfrac{5(26)+4(11)+3(2)+2(3)+1(8)}{50}$
$\dfrac{194}{50} \approx 3.9$

40. K. $D = 0.6 - 0.05M$ The problem tells us that if $M = 1$, $D = 0.55$. We can also reason that if the item arrived in December, it would have been there zero months and have no discount. If $M = 12$, $D = 0$. Answer **K** is the only equation that works for both points.

41. A. 22.37 mpg   Plug in $m = 2000$.
$e = -0.01089(2000)+44.15 = 22.37$

42. H. 509   Because the window is half a circle, $A = \dfrac{1}{2}\pi r^2$. The diameter is 36 in, so the radius is 18 in. $A = \dfrac{1}{2}\pi(18)^2 = 162\pi$
$162\pi \approx 509$

43. D. $(x+3)^2+(y-2)^2 = 36$ The equation of a circle is $(x-x_1)^2+(y-y_1)^2 = r^2$ where (x_1, y_1) are the coordinates of the center and r is the radius. Plug in to get the equation $(x-(-3))^2+(y-2)^2 = 6^2$.

44. G. A square with a side length of 19 feet. The fence plus the gate gives a perimeter of $72 + 4 = 76$. To have the largest area, we need a square with side length $76 \div 4 = 19$.

45. C. $\dfrac{5}{2}$ First, solve for a. To turn 25 into 5, we need a square root, so $a = \dfrac{1}{2}$.
Alternately, $(5^2)^a = 5^1$ so $2a = 1$ and $a = \dfrac{1}{2}$. Then solve for b. $2^{a+b} = 2^3$ plug in for a and simplify. $\dfrac{1}{2}+b = 3$ and $b = \dfrac{5}{2}$.

46. J. 16   16 can be divided by 1, 2, 4, 8, and 16.

47. D. $\dfrac{4}{5} \leq x < 1$ $\dfrac{2}{5} - \dfrac{-1}{2} = \dfrac{9}{10}$ $\dfrac{8}{10} \leq \dfrac{9}{10} < \dfrac{10}{10}$
so $\dfrac{4}{5} \leq \dfrac{9}{10} < 1$

48. H. 8   There are 15 values between the first term and the sixteenth. $-3+15x = 117$ Solve to get $x = 8$.

49. A. 0   8 has three factors, $2 \times 2 \times 2$. 30 has three factors, $2 \times 3 \times 5$. However, none of the numbers between 30 and 40 have three factors.

50. G. 3   Find the duration of each trip from the Tour Information chart. Josephine's trip covered 9 miles in 45 min, or 0.75 hours. Her speed was $9 \div 0.75 = 12$ mph. Philip's trip covered 22.5 miles in 90 min, or 1.5 hours. His speed was $22.5 \div 1.5 = 15$ mph. $15 - 12 = 3$

51. E. 41   Set up a system of three equations.
$$100A+125B+150C = 9400$$
$$A+B+C = 71$$
$$A = 2B$$
Substitute $2B$ for A in the first two equations.
$$100(2B)+125B+150C = 9400$$
$$2B+B+C = 71$$
Simplify.
$$325B+150C = 9400$$
$$3B+C = 71$$
Solve the second equation for C.
$$C = 71-3B$$
Plug into the first equation and solve.
$$325B+150(71-3B) = 9400$$
$$325B+10650-450B = 9400$$
$$-125B+10650 = 9400$$
$$-125B = -1250$$
$$B = 10$$

Plug in to find C.
$C = 71 - 3B = 71 - 3(10) = 4$

52. J. 21   Use the Swamp Boat Information chart to find that the length of Bubba's boat is 19 feet. $\tan 48 = \dfrac{opposite}{adjacent} = \dfrac{l}{19}$

$l = 19 \tan 48 \approx 21$

53. D. $\begin{array}{l} y \geq x+1 \\ x^2 + y^2 \leq 4 \end{array}$ The line $y = -x - 1$ has no impact on the shaded region because there is shading on both sides of it. Therefore, it should not be in the answer. The shading is above $y = x + 1$, so it needs a greater than symbol. The shading is inside the circle $x^2 + y^2 = 4$, so it needs a less than sign.

54. F. $\dfrac{120 + \left(\dfrac{120 \sin 34°}{\sin 99°} \right) + \left(\dfrac{120 \sin 47°}{\sin 99°} \right)}{6}$

The law of sines is $\dfrac{a}{\sin A} = \dfrac{b}{\sin B} = \dfrac{c}{\sin C}$.
Find the third angle and plug in all values.
$\dfrac{120}{\sin 99} = \dfrac{b}{\sin 34} = \dfrac{c}{\sin 47}$ Solve for the missing sides. $b = \dfrac{120 \sin 34}{\sin 99}$ and
$c = \dfrac{120 \sin 47}{\sin 99}$. The perimeter of the

triangle is $120 + \dfrac{120 \sin 34}{\sin 99} + \dfrac{120 \sin 47}{\sin 99}$.

Finally, to find the number of trees needed, divide the perimeter by 6.

55. A. 0.075 $P(A \text{ and } B) = P(A) \times P(B)$
$0.30 \times 0.25 = 0.075$

56. J. 4   First, find $f(-2) = 1$ from the graph.
$g(-2) = 5 - f(-2) = 5 - 1 = 4$

57. B. $\dfrac{21}{5}$ Find numbers that work for a, b,

and c. If $a = 4$ and $b = 1$, then c would be a fraction. If $c = 1$ and $b = 15$. Solve to find $a = 4(15) = 60$. Plug in and simplify.
$\dfrac{2(60) + 6(15)}{3(15) + 5(1)} = \dfrac{210}{50} = \dfrac{21}{5}$

58. J. 4   To find a y-intercept, plug in $x = 0$. This is part of the domain of the middle segment because $-5 \leq 0 \leq 6$. $3(0) + 4 = 4$.

59. B. 70   To find the median, list the numbers in order: 20, 20, 20, 30, 35, 40, 40, 55, 55.

The middle number is 35. To find the mean, add all the numbers and divide by 9 to get 35. The sum is $35 + 35 = 70$

60. K. $\begin{bmatrix} (2a+3c) & (2b+3d) \\ (4a+5c) & (4b+5d) \end{bmatrix}$ To find each

term of the answer, multiply the terms in the row of the first matrix by the terms in the column of the second and add the results. Start with the first row of the first matrix, [2 3], and the first column of the

second matrix, $\begin{bmatrix} a \\ c \end{bmatrix}$. Multiply the terms and

add the results: $(2a + 3c)$. Answer **K** is the only one with this in the first row and first column.